普通高等教育物联网工程专业系列教材
新形态立体化教材(有电子课件与音频资源)

物联网专业英语教程

(第二版)

张强华　杨立春　陆巧儿　司爱侠　编著

西安电子科技大学出版社

内 容 简 介

本书是物联网专业英语教材，选材广泛，覆盖通信技术、网络技术、软件与硬件等各个方面，同时兼顾了相关的发展热点。本书内容包括物联网基础、体系与技术、标准与协议、因特网及其设备、泛在网与虚拟专用网络、条形码技术、RFID、WiFi 与蓝牙、无线传感网络及应用、Network、云计算、智慧城市、大数据等。

本书力求体例创新，适合教学。每一单元包含以下部分：课文——包括选材广泛、风格多样、切合实际的两篇专业文章；单词——给出课文中出现的新词，读者由此可以积累物联网专业的基本词汇；词组——给出课文中的常用词组；缩略语——给出课文中出现的、业内人士必须掌握的缩略语；习题——既有针对课文的练习，也有一些开放性的练习；短文翻译——培养读者的翻译能力；参考译文——让读者对照理解和提高翻译能力；难点脚注——即时讲解，注释宽广，具有开放性。附录Ⅰ的"自测题"可以检查学习效果；附录Ⅱ的"物联网英语词汇的构成与翻译"揭示新词的构成方法，提供翻译技巧，对读者"破译"新词大有裨益；附录Ⅲ的"词汇总表"既可用于复习和背诵，也可作为小词典供长期查阅。

本书吸纳了作者 20 多年的 IT 行业英语翻译与图书编写经验，与课堂教学的各个环节紧密切合，支持备课、教学、复习及考试各个教学环节，有配套的 PPT、音频资源、参考答案、参考试卷等，读者可扫描二维码获取这些资源。

本书既可作为高等院校物联网相关专业的专业英语教材，也可供从业人员自学，还可作为培训班教材使用。

图书在版编目(CIP)数

物联网专业英语教程 / 张强华等编著. —2 版. —西安：西安电子科技大学出版社, 2019.7(2022.10 重印)
ISBN 978-7-5606-4469-1

Ⅰ. ①物… Ⅱ. ①张… Ⅲ. ①互联网络—应用—英语—教材 ②智能技术—应用—英语—教材 Ⅳ. ①TP393.4 ②TP18

中国版本图书馆 CIP 数据核字(2019)第 125350 号

策　　划　李惠萍　毛红兵
责任编辑　雷鸿俊
出版发行　西安电子科技大学出版社(西安市太白南路 2 号)
电　　话　(029)88202421　88201467　　邮　编　710071
网　　址　www.xduph.com　　　　　　　电子邮箱　xdupfxb001@163.com
经　　销　新华书店
印刷单位　陕西天意印务有限责任公司
版　　次　2019 年 7 月第 2 版　　2022 年 10 月第 7 次印刷
开　　本　787 毫米×1092 毫米　1/16　印 张　18.5
字　　数　429 千字
印　　数　16 001～19 000 册
定　　价　41.00 元
ISBN 978-7-5606-4469-1/TP
XDUP　4761002-7
如有印装问题可调换

普通高等教育物联网工程专业系列教材
编审专家委员会名单

总顾问： 姚建铨　天津大学、中国科学院院士　教授
顾　问： 王新霞　中国电子学会物联网专家委员会秘书长
主　任： 王志良　北京科技大学信息工程学院首席教授
副主任： 孙小菡　东南大学电子科学与工程学院　教授
　　　　　曾宪武　青岛科技大学信息科学技术学院物联网系主任　教授
委　员：（成员按姓氏笔画排列）
　　　　　王洪君　山东大学信息科学与工程学院副院长　教授
　　　　　王春枝　湖北工业大学计算机学院院长　教授
　　　　　王宜怀　苏州大学计算机科学与技术学院　教授
　　　　　白秋果　东北大学秦皇岛分校计算机与通信工程学院院长　教授
　　　　　孙知信　南京邮电大学物联网学院副院长　教授
　　　　　朱昌平　河海大学计算机与信息学院副院长　教授
　　　　　邢建平　山东大学电工电子中心副主任　教授
　　　　　刘国柱　青岛科技大学信息科学技术副院长　教授
　　　　　张小平　陕西物联网实验研究中心主任　教授
　　　　　张　申　中国矿业大学物联网中心副主任　教授
　　　　　李仁发　湖南大学教务处处长　教授
　　　　　李朱峰　北京师范大学物联网与嵌入式系统研究中心主任　教授
　　　　　李克清　常熟理工学院计算机科学与工程学院副院长　教授
　　　　　林水生　电子科技大学通信与信息工程学院物联网工程系主任　教授

赵付青	兰州理工大学计算机与通信学院副院长 教授
赵庶旭	兰州交通大学电信工程学院计算机科学与技术系副主任 教授
武奇生	长安大学电子与控制工程学院交通信息与控制系主任 教授
房　胜	山东科技大学信息科学与工程学院物联网专业系主任 教授
施云波	哈尔滨理工大学测控技术与通信学院传感网技术系主任 教授
桂小林	西安交通大学网络与可信计算技术研究中心主任 教授
秦成德	西安邮电大学教学督导 教授
黄传河	武汉大学计算机学院副院长 教授
黄　炜	电子科技大学通信与信息工程学院 教授
黄贤英	重庆理工大学计算机科学与技术系主任 教授
彭　力	江南大学物联网系副主任 教授
谢红薇	太原理工大学计算机科学与技术学院系主任 教授
薛建彬	兰州理工大学计算机与通信学院系主任 副教授

项目策划：毛红兵
策　　划：张　媛　邵汉平　刘玉芳　王　飞

****** 前　　言 ******

物联网是继互联网之后的又一次技术革命，它把网络延伸到物的层面。物联网比互联网更有增长潜力，有可能成为信息产业中继计算机、互联网之后的第三次浪潮。我国物联网已经进入高速发展期，许多高校开设了物联网专业，培养急需的专业人员。由于物联网有极高的发展速度，从业人员必须掌握许多新技术、新方法，因此对专业英语要求较高。具备相关职业技能并精通外语的人员往往会赢得竞争优势，成为职场中不可或缺的核心人才与领军人物。

本书的特点与优势如下：

(1) 选材全面，包括通信技术、网络技术、软件与硬件等各个方面，同时兼顾相关的发展热点。书中许多内容非常实用，具有广阔的覆盖面。作者对丰富的课文素材进行了严谨推敲与细致加工，使其具有教材特性。

(2) 内容全面，包括物联网基础、体系与技术、标准与协议，因特网及其设备，泛在网与虚拟专用网络，条形码技术，RFID，WiFi与蓝牙，无线传感网络及应用，Network，云计算与物联网未来，智慧城市，大数据等。

(3) 体例创新，非常适合教学，与课堂教学的各个环节紧密切合，支持备课、教学、复习及考试各个教学环节。每个单元均包含以下部分：课文——包含选材广泛、风格多样、切合实际的两篇专业文章；单词——给出课文中出现的新词，读者由此可以积累物联网专业的基本词汇；词组——给出课文中的常用词组；缩略语——给出课文中出现的、业内人士必须掌握的缩略语；习题——既有针对课文的练习，也有一些开放性的练习；短文翻译——培养读者的翻译能力；参考译文——让读者对照理解和提高翻译能力；难点脚注——即时讲解，注释宽广，具有开放性，是对课文的延伸与扩展。

(4) 习题量适当，题型丰富，难易搭配，便于教师组织教学。

(5) 附录 I 的"自测题"可以检查学习效果；附录 II 的"物联网英语词汇的构成与翻译"揭示新词的构成方法，提供翻译技巧，对读者"破译"新词大有裨益；附录 III 的"词汇总表"既可用于复习和背诵，也可作为小词典供长期查阅。

(6) 教学支持完善，有配套的 PPT、音频资源、参考答案、参考试卷等，读者可扫描二维码获取这些资源。

(7) 作者有20多年IT行业英语图书的编写经验。在作者编写的英语书籍中，有三部国家级"十一五"规划教材及一部全国畅销书，一部教材获华东地区教材二等奖。这些编写经验有助于本书的完善与提升。

在使用本书的过程中，有任何问题都可以通过电子邮件与我们交流，我们一定会给予答复。邮件标题请注明姓名及"索取物联网英语参考资料"字样。我们的E-mail地址为zqh3882355@sina.com和zqh3882355@163.com。

如本书有任何不妥之处，望大家不吝赐教，让我们共同努力，使本书成为一部"符合学生实际、切合行业实况、知识实用丰富、严谨开放创新"的优秀教材。

作　者

2017年2月

****** 第一版前言 ******

物联网是继互联网之后的又一次技术革命，它把网络延伸到物的层面。物联网比互联网更有增长潜力，有可能成为信息产业中继计算机、互联网之后的第三次浪潮。我国物联网已经进入高速发展期，许多高校开设了物联网专业，培养急需的专业人员。由于物联网有极高的发展速度，从业人员必须掌握许多新技术、新方法，因此对专业英语要求较高。具备相关职业技能并精通外语的人员往往会赢得竞争优势，成为职场中不可或缺的核心人才与领军人物。

本书的特点与优势如下：

(1) 选材全面，包括通信技术、网络技术、软件与硬件等各个方面，同时兼顾相关的发展热点。书中许多内容非常实用，具有广阔的覆盖面。作者对丰富的课文素材进行了严谨推敲与细致加工，使其具有教材特性。

(2) 内容全面，包括物联网基础、体系与技术、标准与协议，因特网及其设备，泛在网与虚拟专用网络，条形码技术，RFID 与 EPC 网络，WiFi 与蓝牙，无线传感网络与应用，云计算与物联网未来等。

(3) 体例创新，非常适合教学，与课堂教学的各个环节紧密切合，支持备课、教学、复习及考试各个教学环节。每个单元均包含以下部分：课文——选材广泛、风格多样、切合实际的两篇专业文章；单词——给出课文中出现的新词，读者由此可以积累物联网专业的基本词汇；词组——给出课文中的常用词组；缩略语——给出课文中出现的、业内人士必须掌握的缩略语；习题——既有针对课文的练习，也有一些开放性的练习；短文翻译——培养读者的翻译能力；参考译文——让读者对照理解和提高翻译能力；难点脚注——即时讲解，注释宽广，具有开放性，是对课文的延伸与扩展。

(4) 习题量适当，题型丰富，难易搭配，便于教师组织教学。

(5) 附录 I 的"自测题"可以检查学习效果；附录 II 的"物联网英语词汇的构成与翻译"揭示新词的构成方法，提供翻译技巧，对读者"破译"新词大有裨益；附录 III 的"词汇总表"既可用于复习和背诵，也可作为小词典供长期查阅。

(6) 教学支持完善，有配套的 PPT、参考答案、参考试卷等。

(7) 作者有近 20 年 IT 行业英语图书的编写经验。在作者编写的英语书籍中，有三部国家级"十一五"规划教材，一部全国畅销书，一部获华东地区教材二等奖图书。这些编写经验有助于本书的完善与提升。

在使用本书的过程中，有任何问题都可以通过电子邮件与我们交流，我们一定会给予答复。邮件标题请注明姓名及"索取物联网英语参考资料"字样。我们的 E-mail 地址为 zqh3882355@sina.com 和 zqh3882355@163.com。

如本书有任何不妥之处，望大家不吝赐教，让我们共同努力，使本书成为一部"符合学生实际、切合行业实况、知识实用丰富、严谨开放创新"的优秀教材。

作　者
2012 年 10 月

目 录

Unit 1	Internet of Things
[1] Text A	Internet of Things
[6] New Words	
[9] Phrases	
[10] Abbreviations	
[10] Exercises	
[13] Text B	Applications of IoT
[16] New Words	
[18] Phrases	
[19] Abbreviations	
[19] Exercises	
[19] Text A 参考译文	物联网

Unit 2	Internet
[23] Text A	Internet
[26] New Words	
[27] Phrases	
[28] Abbreviations	
[28] Exercises	
[30] Text B	Network Device
[34] New Words	
[35] Phrases	
[36] Abbreviations	
[36] Exercises	
[37] Text A 参考译文	因特网

Unit 3	Architecture and Technology of IoT
[40] Text A	Architecture, Hardware, Software and Algorithms of IoT
[43] New Words	
[44] Phrases	
[45] Abbreviations	
[46] Exercises	

[48]	Text B	Technology of IoT
[52]	New Words	
[54]	Phrases	
[54]	Abbreviations	
[54]	Exercises	
[55]	Text A 参考译文	物联网的体系、硬件、软件及算法

Unit 4		**Ubiquitous Network and VPNs**
[58]	Text A	How Ubiquitous Networking Will Work?
[61]	New Words	
[63]	Phrases	
[63]	Abbreviations	
[63]	Exercises	
[65]	Text B	VPNs
[71]	New Words	
[72]	Phrases	
[73]	Abbreviations	
[73]	Exercises	
[74]	Text A 参考译文	泛在网是如何工作的？

Unit 5		**Barcode**
[77]	Text A	Barcode
[81]	New Words	
[83]	Phrases	
[84]	Abbreviations	
[84]	Exercises	
[86]	Text B	How 2D Bar Codes will Work?
[90]	New Words	
[92]	Phrases	
[93]	Abbreviations	
[93]	Exercises	
[94]	Text A 参考译文	条形码

Unit 6		**Radio Frequency Identification**
[97]	Text A	RFID Basic
[100]	New Words	
[102]	Phrases	
[102]	Abbreviations	

[102]	Exercises	
[105]	Text B	How RFID Works?
[111]	New Words	
[112]	Phrases	
[113]	Abbreviations	
[113]	Exercises	
[114]	Text A 参考译文	RFID 基础

Unit 7		WiFi and Bluetooth
[117]	Text A	How WiFi Works?
[122]	New Words	
[123]	Phrases	
[123]	Abbreviations	
[124]	Exercises	
[125]	Text B	Bluetooth
[130]	New Words	
[131]	Phrases	
[132]	Abbreviations	
[132]	Exercises	
[133]	Text A 参考译文	WiFi 是如何工作的？

Unit 8		Wireless Sensor Network and It's Application
[137]	Text A	Wireless Sensor Network
[142]	New Words	
[144]	Phrases	
[145]	Abbreviations	
[145]	Exercises	
[148]	Text B	The Application of Wireless Sensor Network
[149]	New Words	
[150]	Phrases	
[151]	Exercises	
[152]	Text A 参考译文	无线传感器网络

Unit 9		Network
[156]	Text A	IEEE 802.15.4
[163]	New Words	
[164]	Phrases	
[165]	Abbreviations	

[165]	Exercises	
[167]	Text B	ZigBee
[171]	New Words	
[171]	Phrases	
[172]	Abbreviations	
[172]	Exercises	
[173]	Text A 参考译文	IEEE 802.15.4

Unit 10		Cloud Computing
[177]	Text A	How Cloud Computing Works?
[181]	New Words	
[182]	Phrases	
[183]	Exercises	
[185]	Text B	4G—The Future of Mobile Internet
[187]	New Words	
[188]	Phrases	
[189]	Abbreviations	
[189]	Exercises	
[190]	Text A 参考译文	云计算是如何工作的？

Unit 11		Smart City
[193]	Text A	Smart City
[197]	New Words	
[199]	Phrases	
[200]	Abbreviations	
[200]	Exercises	
[203]	Text B	The Impact of the Internet of Things on Big Data
[206]	New Words	
[207]	Phrases	
[207]	Abbreviations	
[207]	Exercises	
[208]	Text A 参考译文	智慧城市

[211]	附录 I	自测题
[222]	附录 II	物联网英语词汇的构成与翻译
[225]	附录 III	词汇总表

Unit 1 Internet of Things

Internet of Things

The Internet of Things (IoT) refers to uniquely identifiable objects (things) and their virtual representations in an Internet-like structure. The term Internet of Things was first used by Kevin Ashton[①] in 1999. The concept of the Internet of Things first became popular through the Auto-ID Center[②] and related market analysts publications. Radio-Frequency IDentification (RFID) is often seen as a prerequisite for the Internet of Things. If all objects of daily life were equipped with radio tags, they could be identified and inventoried by computers. However, unique identification of things may be achieved through other means such as barcodes or 2D-codes as well.

With all objects in the world equipped with minuscule identifying devices, daily life on Earth would undergo a transformation. Companies would not run out of stock or waste products, as all involved parties would know exactly which products are required and consumed. Mislaid and stolen items would be easily tracked and located.

1. Alternative definitions

Different definitions for the Internet of Things have appeared and the term is evolving as the technology and implementation of the ideas move forward. Here are several partially overlapping definitions.

1.1 CASAGRAS

A global network infrastructure, linking physical and virtual objects through the exploitation of data capture and communication capabilities. This infrastructure includes existing and

---

① Kevin Ashton (born in 1968 in Birmingham, England) is a British technology pioneer ([ˌpaɪəˈnɪə]n.先驱，倡导者) who cofounded the Auto-ID Center at the Massachusetts Institute of Technology, which created a global standard system for RFID and other sensors.
② Auto-ID Center to design the architecture for the Internet of Things together with EPCglobal.

evolving Internet and network developments. It will offer specific object-identification, sensor and connection capability as the basis for the development of independent cooperative services and applications. These will be characterized by a high degree of autonomous data capture, event transfer, network connectivity and interoperability.

1.2 SAP[①]

A world where physical objects are seamlessly integrated into the information network, and where the physical objects can become active participants in business processes. Services are available to interact with these "smart objects" over the Internet, query and change their state and any information associated with them, taking into account security and privacy issues.

1.3 EPoSS

The network formed by things/objects having identities, virtual personalities operating in smart spaces using intelligent interfaces to connect and communicate with the users, social and environmental contexts.

1.4 CERP-IoT

Internet of Things is an integrated part of Future Internet. It could be defined as a dynamic global network infrastructure with self configuring capabilities based on standard and interoperable communication protocols. In the IoT, physical and virtual "things" have identities, physical attributes, and virtual personalities and use intelligent interfaces, and are seamlessly integrated into the information network. In the IoT, "things" are expected to become active participants in business, information and social processes. They are enabled to interact and communicate among themselves and with the environment by exchanging data and information "sensed" about the environment, while reacting autonomously to the "real/physical world" events and influencing it by running processes that trigger actions and create services with or without direct human intervention. Interfaces in the form of services facilitate interactions with these "smart things" over the Internet, query and change their state and any information associated with them, taking into account security and privacy issues.

1.5 Other

The future Internet of Things links uniquely identifiable things to their virtual representations in the Internet containing or linking to additional information on their identity, status, location or any other business, social or privately relevant information at a financial or non-financial pay-off. It exceeds the efforts of information provisioning and offers information access to non-predefined participants. The provided accurate and appropriate information may be accessed in the right quantity and condition, at the right time and place at the right price. The Internet of Things is not synonymous with ubiquitous/pervasive computing, the Internet Protocol (IP), communication technology, embedded devices, its applications, the Internet of People or the Intranet/Extranet of Things, yet it relies on all of these approaches. The

---

① SAP is a German software corporation that makes enterprise software to manage business operations and customer relations.

association of intelligent virtual representations (e.g. called avatars and embedded, hosted in the Cloud or centralized) and physical objects are sometimes called "cyberobjects". Cyberobjects are then considered as autonomous actors of the value chains they are involved in: able to perceive, analyze and react in various contexts; although acting under the guidance of human beings as programmed. Cyberobjects can then be assistants, advisors, decision makers, etc; and can be considered as true agent (economics)①, helping to change existing economic or organization models. In such a scenario, the conception of avatars refers to Artifical Intelligence② and complex system.

2. Unique addressability of things

The original idea of the Auto-ID Center is based on RFID-tags and unique identification through the Electronic Product Code③.

An alternative view, from the world of the Semantic Web④, focuses instead on making all things (not just those electronic, smart, or RFID-enabled) addressable by the existing naming protocols, such as URI⑤. The objects themselves do not converse, but they may now be referred to by other agents, such as powerful centralized servers acting for their human owners.

The next generation of Internet applications using Internet Protocol version 6 (IPv6)⑥ would be able to communicate with devices attached to virtually all human-made objects because of the extremely large address space of IPv6. This system would therefore be able to identify any kind of object.

A combination of these ideas can be found in the current GS1/EPCglobal⑦ EPC Information Services specifications. This system is being used to identify objects in industries ranging from Aerospace to Fast Moving Consumer Products and Transportation Logistics.

---

① In economics, an agent is an actor and decision maker in a model. Typically, every agent makes decisions by solving a well or ill defined optimization/choice problem.
② Artificial Intelligence (AI) is the intelligence of machines and the branch of computer science that aims to create it. AI textbooks define the field as "the study and design of intelligent agents" where an intelligent agent is a system that perceives its environment and takes actions that <u>maximize</u> (['mæksmaiz]vt.取……最大值,最佳化) its chances of success.
③ The Electronic Product Code (EPC, 电子产品编码) is designed as a universal identifier that provides a unique identity for every physical object anywhere in the world, for all time.
④ The Semantic Web is a collaborative movement led by the World Wide Web(WWW, 万维网) Consortium (W3C) that promotes common formats for data on the World Wide Web. By encouraging the inclusion of semantic content in web pages, the Semantic Web aims at converting the current web of unstructured documents into a "web of data".
⑤ In computing, a Uniform Resource Identifier (URI) is a string of characters used to identify a name or a resource on the Internet.
⑥ Internet Protocol version 6 (IPv6) is a version of the Internet Protocol (IP). It is designed to <u>succeed</u> ([sək'si:d]v.继……之后) the Internet Protocol version 4 (IPv4). The Internet operates by transferring data between hosts in small packets that are independently routed across networks as specified by an international communications protocol known as the Internet Protocol.
⑦ EPCglobal is a joint venture between GS1 (formerly known as EAN International) and GS1 US (formerly the Uniform Code Council, Inc.). It is an organization <u>set up</u> (建立) to achieve worldwide adoption and standardization of EPC technology.

3. Trends and characteristics

3.1 Intelligence

Ambient Intelligence① and Autonomous Control are not part of the original concept of the Internet of Things. Ambient Intelligence and Autonomous Control do not necessarily require Internet structures, either. However, there is a shift in research to integrate the concepts of the Internet of Things and Autonomous Control. In the future the Internet of Things may be a nondeterministic and open network in which auto-organized or intelligent entities (Web services②, SOA③ components), virtual objects will be interoperable and able to act independently (pursuing their own objectives or shared ones) depending on the context, circumstances or environments.

Embedded intelligence presents an "AI-oriented" perspective of IoT, which can be more clearly defined as: leveraging the capacity to collect and analyze the digital traces left by people when interacting with widely deployed smart things to discover the knowledge about human life, environment interaction, as well as social connection/behavior.

3.2 Architecture

The system will likely be an example of Event-Driven Architecture④, bottom-up made (based on the context of processes and operations, in real-time) and will consider any subsidiary level. Therefore, model driven and functional approaches will coexist with new ones able to treat exceptions and unusual evolution of processes.

3.3 Complex system

In semi-open or closed loops, it will therefore be considered and studied as a Complex system due to the huge number of different links and interactions between autonomous actors, and its capacity to integrate new actors. At the overall stage (full open loop) it will likely be seen as a chaotic⑤ environment.

---

① In computing, Ambient Intelligence (AmI, 环境智能) refers to electronic environments that are sensitive and responsive to the presence of people. In an Ambient Intelligence world, devices work in concert to support people in carrying out their everyday life activities, tasks and rituals in easy, natural way using information and intelligence that is hidden in the network connecting these devices.

② A Web service is a method of communication between two electronic devices over the web. The W3C defines a "Web service" as "a software system designed to support interoperable machine-to-machine (机器对机器) interaction over a network".

③ In software engineering, a Service-Oriented Architecture (SOA) is a set of principles and methodologies ([meθəˈdɔlədʒi]n.方法学, 方法论) for designing and developing software in the form of interoperable services. These services are well-defined business functionalities that are built as software components (discrete pieces of code (代码段) and/or data structures (数据结构)) that can be reused for different purposes. SOA design principles are used during the phases of systems development and integration.

④ Event-Driven Architecture (EDA) is a software architecture pattern promoting the production, detection, consumption of, and reaction to events.

⑤ Chaos theory (混沌论) is a field of study in mathematics, with applications in several disciplines including physics, economics, biology, and philosophy. Chaos theory studies the behavior of dynamical systems that are highly sensitive to initial conditions, an effect which is popularly referred to as the butterfly effect (蝴蝶效应).

3.4 Size considerations

The Internet of Objects would encode 50 to 100 trillion objects, and be able to follow the movement of those objects.

3.5 Time considerations

In this Internet of Things, made of billions of parallel and simultaneous events, time will no more be used as a common and linear dimension but will depend on each entity (object, process, information system, etc.). This Internet of Things will be accordingly based on massive parallel IT systems (Parallel computing①).

3.6 Space considerations

In an Internet of Things, the precise geographic location of a thing — and also the precise geographic dimensions of a thing — will be critical. Currently, the Internet has been primarily used to manage information processed by people. Therefore, facts about a thing, such as its location in time and space, has been less critical to track because the person processing the information can decide whether or not that information was important to the action being taken, and if so, add the missing information (or decide not to take the action). (Note that some things in the Internet of Things will be sensors, and sensor location is usually important.) The GeoWeb② and Digital Earth③ are promising applications that become possible when things can become organized and connected by location. However, challenges that remain include the constraints of variable spatial scales, the need to handle massive amounts of data, and an indexing for fast search and neighbour operations. In the Internet of Things, if things are able to take actions on their own initiative, this human-centric mediation role is eliminated, and the time-space context that we as humans take for granted must be given a central role in this information ecosystem. Just as standards play a key role in the Internet and the Web, geospatial standards will play a key role in the Internet of Things.

4. Frameworks

Internet of Things frameworks might help support the interaction between "things" and allow for more complex structures like distributed computing④ and the development of distributed applications. Currently, Internet of Things frameworks seem to focus on real time data

---

① Parallel computing is a form of computation in which many calculations are carried out simultaneously, operating on the principle that large problems can often be divided into smaller ones, which are then solved concurrently ([kənˈkʌrəntli]adj.并发的) ("in parallel").

② The Geospatial Web or Geoweb is a relatively new term that implies the merging of geographical (location-based) information with the abstract information that currently dominates the Internet. This would create an environment where one could search for things based on location instead of by keyword only – e.g. "What is here?"

③ Digital Earth is the name given to a concept by former US vice president Al Gore in 1998, describing a virtual representation of the Earth that is spatially ([ˈspeiʃəli]adv.空间地) referenced and interconnected with the world's digital knowledge archives.

④ Distributed computing is a field of computer science that studies distributed systems. A distributed system consists of (由……组成) multiple autonomous computers that communicate through a computer network.

logging solutions like Pachube[①]: offering some basis to work with many "things" and have them interact. Future developments might lead to specific software development environments[②] to create the software to work with the hardware used in the Internet of Things.

New Words

uniquely	[juːˈniːkli]	adv. 独特地，唯一地
identifiable	[aiˈdentifaiəbl]	adj. 可以确认的
virtual	[ˈvəːtjuəl]	adj. 虚拟的，实质的
representation	[ˌreprizenˈteiʃən]	n. 表示法，表现
popular	[ˈpɔpjulə]	adj. 流行的，受欢迎的
publication	[ˌpʌbliˈkeiʃən]	n. 出版物，出版，发行，发表
prerequisite	[ˈpriːˈrekwizit]	n. 先决条件
		adj. 首要必备的
tag	[tæg]	n. 标签，标识
identify	[aiˈdentifai]	vt. 识别，鉴别
inventory	[ˈinvəntri]	vt. 编制……的目录；盘存，清查
		vi. 对清单上存货的估价
		n. 详细目录，存货，财产清册，总量
barcode	[ˈbɑːkəud]	n. 条形码
2D-code		n. 二维码
minuscule	[miˈnʌskjuːl]	adj. 极小的
undergo	[ˌʌndəˈgəu]	vt. 经历，遭受，忍受
stock	[stɔk]	n. 库存，原料
misland	[misˈlænd]	vt. 放错，遗失
alternative	[ɔːlˈtəːnətiv]	adj. 选择性的，二中择一的
definition	[ˌdefiˈniʃən]	n. 定义，解说，精确度
evolve	[iˈvɔlv]	v. (使)发展，(使)进展，(使)进化
overlapping	[ˈəuvəˈlæpiŋ]	n. 重叠，搭接
infrastructure	[ˈinfrəstrʌktʃə]	n. 基础，下部构造，基础组织
exploitation	[ˌeksplɔiˈteiʃən]	n. 使用
communication	[kəˌmjuːniˈkeiʃn]	n. 通信
capability	[ˌkeipəˈbiliti]	n. (实际)能力，性能，容量，接受力
specific	[spiˈsifik]	adj. 详细而精确的，明确的，特殊的

---

① Pachube is an on-line database service provider allowing developers to connect sensor data to the Web and to build their own applications on it.

② An Integrated Development Environment (IDE，集成开发环境) (also known as integrated design environment, integrated debugging environment or interactive development environment) is a software application that provides comprehensive ([ˌkɔmpriˈhensiv]adj.全面的，广泛的) facilities to computer programmers for software development.

sensor	[ˈsensə]	n. 传感器
independent	[ˌindiˈpendənt]	adj. 独立的
cooperative	[kəuˈɔpərətiv]	adj. 合作的，协力的
application	[ˌæpliˈkeiʃən]	n. 应用，应用程序
participant	[pɑːˈtisipənt]	n. 参与者，共享者
available	[əˈveiləbl]	adj. 可用的，有效的
query	[ˈkwiəri]	v. 询问，查询
state	[steit]	n. 情形，状态
security	[siˈkjuəriti]	n. 安全
privacy	[ˈpraivəsi]	n. 秘密
issue	[ˈisjuː]	n. 问题，结果
identity	[aiˈdentiti]	n. 身份，特性
personality	[ˌpəːsəˈnæliti]	n. 人格，人物，人名
environmental	[inˌvaiərənˈmentl]	adj. 周围的，环境的
dynamic	[daiˈnæmik]	adj. 动态的
configure	[kənˈfigə]	vi. 配置，设定
interoperable	[ˌintərˈɔpərəbl]	adj. 能共同操作的，能共同使用的
attribute	[əˈtribjuː(ː)t]	n. 属性，品质，特征
exchange	[iksˈtʃeindʒ]	vt. & n. 交换，调换，兑换，交易
sense	[sens]	vt. 感知，感到，认识
react	[riˈækt]	vi. 起反应，起作用
influence	[ˈinfluəns]	n. 影响，(电磁)感应
		vt. 影响，改变
interaction	[ˌintərˈækʃən]	n. 交互作用，交感
facilitate	[fəˈsiliteit]	vt. 使便利，帮助，使容易，促进
location	[ləuˈkeiʃən]	n. 位置，场所
relevant	[ˈrelivənt]	adj. 有关的，相应的
predefine	[ˈpriːdiˈfain]	vt. 预先确定
accurate	[ˈækjurit]	adj. 正确的，精确的
access	[ˈækses]	n. & vt. 访问，存取
pervasive	[pəːˈveisiv]	adj. 普遍深入的
embedded	[emˈbedid]	adj. 植入的，嵌入的，内含的
intranet	[ˈintrənet]	n. 内联网
extranet	[ˈekstrənet]	n. 外联网
association	[əˌsəusiˈeiʃən]	n. 协会，联合
avatar	[ˌævəˈtɑː]	n. 化身，天神下凡，具体化
centralize	[ˈsentrəlaiz]	vt. 集聚，集中
host	[həust]	n. 主机
		v. 做主机

cyberobject	['saibəob'dʒikt]	n. 计算体，计算部件
perceive	[pə'siːv]	v. 感知，感到
analyze	['ænəlaiz]	vt. 分析，分解
assistant	[ə'sistənt]	n. 助手，助教
		adj. 辅助的，助理的
advisor	[əd'vaizə]	n. 顾问
scenario	[si'nɑːriəu]	n. 情景，情节
addressability	[ə'dresəbiliti]	n. 可寻址能力
addressable	[ə'dresəbl]	adj. 可设定地址的
converse	[kən'vəːs]	vi. 认识；谈话，交谈
virtually	['veːtjuəli]	adv. 事实上，实质上；几乎
specification	[ˌspesifi'keiʃən]	n. 详述，规格，说明书，规范
original	[ə'ridʒənəl]	adj. 最初的，原始的，独创的，新颖的
integrate	['intigreit]	vt. 使成整体，使一体化
nondeterministic	['nɔnditəːmi'nistik]	adj. 非定常的，非确定的
deploy	[di'plɔi]	v. 展开，配置
behavior	[bi'heivjə]	n. 举止，行为
architecture	['ɑːkitektʃə]	n. 体系结构
subsidiary	[səb'sidjəri]	adj. 辅助的，补充的
functional	['fʌŋkʃənl]	adj. 功能的
approach	[ə'prəutʃ]	n. 方法，步骤，途径
coexist	[kəuig'zist]	vi. 共存
exception	[ik'sepʃən]	n. 除外，例外，反对，异议
syntactic	[sin'tæktik]	adj. 句法，语法
consequently	['kɔnsikwəntli]	adv. 从而，因此
accordingly	[ə'kɔːdiŋli]	adv. 因此，从而
self-referenced	[self-'refərensd]	adj. 自引用的
stage	[steidʒ]	n. 发展的进程、阶段或时期
chaotic	[kei'ɔtik]	adj. 混乱的，无秩序的，混沌的
simultaneous	[ˌsiməl'teinjəs]	adj. 同时的，同时发生的
massive	['mæsiv]	adj. 可观的，巨大的，大量的
precise	[pri'sais]	adj. 精确的，准确的
geographic	[ˌdʒiə'græfik]	adj. 地理学的，地理的
critical	['kritikəl]	adj. 极重要的，至关紧要的
promising	['prɔmisiŋ]	adj. 有希望的，有前途的
constraint	[kən'streint]	n. 约束，强制
variable	['vɛəriəbl]	n. 变数，变量
		adj. 可变的，易变的，变量的

index	['indeks]	n. 索引
		vi. 做索引
mediation	[ˌmiːdi'eiʃən]	n. 仲裁，调停，调解
geospatial	['dʒiːəuspeiʃəl]	adj. 地理空间的
ecosystem	[iːkə'sistəm]	n. 生态系统
framework	['freimwəːk]	n. 构架，框架，结构

Phrases

Internet of Things(IOT)	物联网
refer to	涉及，关系到；提到，谈到
market analyst	市场分析者
radio tag	无线标签，无线标识
waste product	废品，次品
move forward	前进
data capture	数据捕捉
event transfer	事件传输
be integrated into	统一到……中，整合到……中
smart object	智能物体，智能对象
take into account	重视，考虑
intelligent interface	智能接口
smart space	智能空间
be defined as	被定义为
base on	基于
be enabled to	使能够
trigger action	触发作用
Internet Protocol (IP)	因特网协议
rely on	依赖，依靠
be involved in	涉及，专心
value chain	价值链
under the guise of	假借，以……为幌子
decision maker	决策者
Artifical Intelligence (AI)	人工智能
focus on	集中
attach to	把……放在，附加到……
human-made object	人造物体
address space	地址空间
Fast Moving Consumer Product	快速消费品
Ambient Intelligence	环境智能

Autonomous Control	自主控制
Event-Driven Architecture	事件驱动的体系结构
bottom-up made	自底向上的
closed loop	闭环
due to	由于
full open loop	全开环
linear dimension	线性维度
parallel computing	并行计算
Digital Earth	数字地球
spatial scale	空间尺度，空间比例尺，空间等级
on their own initiative	他们自己主动
distributed computing	分布计算
data logging	数据资料记录

Abbreviations

RFID (Radio-Frequency IDentification)	射频识别
CASAGRAS (Coordination and Support Action for Global RFID-related Activities and Standardization)	全球RFID运作及标准化协调支持行动
EPOSS (the European Technology Platform on Smart Systems Integration)	欧洲智能系统集成技术平台
CERT (Cluster of European Research Projects)	欧盟物联网研究项目组
URI (Uniform Resource Identifier)	统一资源标识符
SOA (Service-Oriented Architecture)	面向服务的体系结构

Exercises

I. Answer the following questions according to the text.

1. What does the Internet of Things refer to? Who used the term first and when?
2. What does RFID stand for? What is it often seen as?
3. What is CASAGRAS?
4. What do physical and virtual "things" have in the IoT?
5. What is the original idea of the Auto-ID Center based on?
6. What would the next generation of Internet applications which use Internet Protocol version 6 be able to do? Why?
7. What may the Internet of Things be in the future?
8. What can embedded intelligence be more clearly defined as?
9. What will be critical in an Internet of Things?
10. What do Internet of Things frameworks seem to focus on currently? What about future?

Unit 1　Internet of Things

II. Translate the following terms or phrases from English into Chinese and vice versa.

1. deploy _____ 1. _____
2. capability _____ 2. _____
3. framework _____ 3. _____
4. tag _____ 4. _____
5. n. 索引 vi. 做索引 _____ 5. _____
6. barcode _____ 6. _____
7. identify _____ 7. _____
8. Intranet _____ 8. _____
9. n. 体系结构 _____ 9. _____
10. sensor _____ 10. _____

III. Fill in the blanks with the words given below.

| communications | predicting | devices | development | speed |
| affordable | technology | size | interconnected | seamless |

Internet of Things: How It Will Change the World

Try to imagine a world where everything is ___1___. A world where you can modify your own reality to see what you want to see, where your coffee machine knows when you need your next fix, and the high streets are populated with characters from your favourite PC games.

This future is not some distant dream as imagined by fans of Minority Report. Much of this ___2___ is already here, and the current rate of ___3___ has futurists claiming that this will be a reality within the next five to ten years.

According to a recent report by Amdocs, experts are ___4___ that there will be seven trillion networked devices by 2017, delivering a connected life that has immediate access to data, media, communities and ___5___ across a broad range of ___6___.

We have been promised this interconnected world since the 1980s, but limitations such as costs and the ___7___ and capabilities of chips and infrastructure, have kept many innovations on hold.

However, 4G and IPv6 now offer vast superhighways of space and ___8___, delivering what's needed for Machine to Machine (M2M) communication to take place on a grand scale. Add to this the fact that Moore's Law remains a constant, chips have become both smaller and more ___9___, and that technology is developing at an unprecedented rate, and you have all the ingredients necessary for an explosion in M2M implementations across the globe.

Communication service providers are moving towards Tera-play, evolving to accommodate a complex ecosystem that's fast and intelligent, and constantly changing to meet market demand and innovation. Tera-play providers will do more than just connect billions of people to one another—they will be a part of M2M, ensuring that communications between trillions of devices is ___10___ and organic.

IV. Translate the following passages from English to Chinese.

✧✧✧ Passage One ✧✧✧

In computing, the Internet of Things refers to a network of objects, such as household appliances. It is often a self-configuring wireless network. The concept of the Internet of Things is attributed to the original Auto-ID Center, founded in 1999 and based at the time in Massachusetts Institute of Technology (MIT).

The idea is as simple as its application is difficult. If all cans, books, shoes or parts of cars are equipped with minuscule identifying devices, daily life on our planet will undergo a transformation. Things like running out of stock or wasted products will no longer exist as we will know exactly what is being consumed on the other side of the globe. Theft will be a thing of the past as we will know where a product is at all times. The same applies to parcels lost in the post.

If all objects of daily life, from yogurt to an airplane, are equipped with radio tags, they can be identified and managed by computers in the same way humans can. The next generation of Internet applications (IPv6 protocol) would be able to identify more objects than IPv4, which is currently in use. This system would therefore be able to instantaneously identify any kind of object.

The Internet of Objects should encode 50 to 100,000 billion objects and follow the movement of those objects. Every human being is surrounded by 1,000 to 5,000 objects.

✧✧✧ Passage Two ✧✧✧

The Internet of Things—This is Where we're Going

In one vision of the future, every "thing" is connected to the Internet. This "Internet of Things" will bring about revolutionary change in how we interact with our environment and, more importantly, how we live our lives.

The idea of everything being connected to the Internet is not new, but it's increasingly becoming a reality. The Internet of Things came into being in 2008 when the number of things connected to the Internet was greater than the number of people who were connected.

The technical utopians have portrayed the Internet of Things as a good thing that will bring untold benefits. They are supported by all the companies that stand to benefit by the increasing connectedness of everything.

Universal connectivity, sensors and computers that are able to collect, analyse and act on this data will bring about improvements in health, food production. In a roundabout way, it might even alleviate poverty.

On the other side are the sceptics who warn of the dangers inherent in not only having an ever growing Internet of Things, but our increasing reliance on it.

Text B

Applications of IoT

1. Retail

The first large scale application of the Internet of Things technologies will be to replace the bar code in retail. The main barriers so far have been the much higher cost of the tag over the bar code, some needed technology improvement for what concerns transmission of metals and liquid items, and privacy concerns. Nonetheless, the replacement has already started in some pilot projects and although one may expect to see coexistence of the two identification mechanisms for many years into the future, advances in the electronics industry will render the RFID tag ever cheaper and more attractive and accessible to the retailers.

The electronic tags offer multiple benefits over the bar code for both the retailers and the consumers. The retailers will have item identification unified from the producer, through the storage, the shop floor, cashier and check out, as well as theft protection. They may also save cost by allowing customers to check out the products themselves and without having to put the bought items on a conveyer belt. The shelves may be intelligent issuing a refill order automatically to the storage as items are sold offering precise delivery from the wholesaler directly to the shelf. Furthermore, the history of any item from production to the shelf can be stored offering increased quality management along the supply chain.

For the consumers this offers the possibility of avoiding long check-out lines and having the product history available, which will improve food safety and protect consumer rights in case of failing products. Yet, RFID in retail has created consumer concerns, although it's hard to see how an electronic tag may infringe privacy any more than the bar code. Any item paid with a payment card in somebody's name may be connected to the owner in the shop's database. The only difference is that the electronic tag could be read post-sale to identify date and location of the purchase. On the other hand, this could be used to prove rightful ownership and sort out guarantee disputes. The challenge is to put this into a useful context for the user and to provide the right incentives to increase acceptance. Similar to the way security equipment in cars gives a discount on insurance, having the capital goods in a household marked with electronic tags making it more difficult to sell the items illegally and easier to recover stolen goods could give discounts on the house insurance. Similar to the way one register the sale of a car among private persons, online access to the items registry could allow lawful transactions to be stored in the item's history.

Today almost every phone sold is equipped with some kind of short range radio communication like Bluetooth, or more specifically Near Field Communication (NFC)[①] specifically designed for reading RFID tags. Predictions indicate that there could be as many as 2 billion NFC enabled mobile phones by 2012. Soon the consumer will no longer need to consult a shop floor reader to know the history of a product, and the shopping list can be created as the wrapping of used goods are discarded. This opens for automated warehouses where the shopping list is transmitted when the customer leaves the house to collect a ready made shopping bag already checked upon arrival to the warehouse. With the ability of directly reading the tags, the inventory of your belongings may be stored in you mobile phone, which will make insurance claims easier and facilitate the private sales of goods since a centralised registry of things will no longer be needed.

2. Logistics

It is important to remember that innovation in logistics normally does not change the industry fundamentally but allows improving efficiency of processes or enables new value adding features. The first observation to be made from the preceding discussion is that the warehouses will become completely automatic with items being checked in and out and orders automatically passed to the suppliers. This will allow better asset management and proactive planning on behalf of the transporter. Goods may be transported without human intervention from producer to consumer and the manufacturers will have a direct feedback on the market's needs. In this way the production and transportation can be adapted dynamically thus saving time, energy, and the environment.

The executable codes in the tags enable the *thing* in transit to make intelligent decisions on its routing based on information received either via readers or positioning systems. This will help optimizing the forwarding of the item and delegate routing authority from the transporter to the manufacturer or the customer. The thing could check back with the sender if it should continue towards the intended recipient, or alternatively move to another recipient paying better to have the thing quickly.

Present day logistics is based on established supply chains from manufacturer to consumer. Supply chains are based on legal agreement and existing over time. It is possible to envision that the things in transit form a marketplace and that a consumer could place a request on the Internet of Things, receive and accept an offer from a thing fulfilling the request. Equivalent to service composition in the virtual software world where an application is assembled of multiple services available on the Internet, may an assembled thing be constructed from parts automatically

---

① Near Field Communication (NFC) is a set of standards for smartphones and similar devices to establish radio communication with each other by touching them together or bringing them into close proximity, usually no more than a few centimetres (['senti,mi:tə]n. 厘米，公分). Present and anticipated applications include contactless transactions, data exchange, and simplified setup of more complex communications such as WiFi. Communication is also possible between an NFC device and an unpowered NFC chip, called a "tag".

identified on the Internet of Things. This will change the way business deals are made since a customer may not place an order for a large volume of things with a manufacturer, but buy them in a sequence of individual orders and possibly from competing manufacturers.

3. Pharmaceutical

Pharmaceutical applications are fundamentally nothing but production, logistics, and retail of drugs as already outlined in the above sections. An added benefit of an electronic tag is that it may carry information related to drug use making it easier for the customer to be acquainted with adverse effects and optimal dosage.

Furthermore, smart biodegradable dust embedded inside pills may interact with the intelligent tag on the box allowing the latter to monitor the use and abuse of medicine and inform the pharmacist when new supply is needed. The smart dust① in pills could know incompatible drugs, and when one is detected close enough the pill could refuse to activate or release the active substances. The same mechanism could of course be used to prevent overdoses. If there is an accident or when someone perishes from drug abuse or misuse it will be possible to quickly identify the taken drug by asking the smart dust, which may also inform about the right antidote and dosage to enable the emergency treatment to be given faster and more correct and thereby saving lives.

4. Food

Europe is traditionally spoiled with excellent food and wine where the quest for the perfect taste has been ongoing for centuries. French law pioneered the idea of protecting produce of a limited geographical origin, and similar laws have since been established in many European countries. Traceable identities will help the consumers to verify the origins of the products and help Europe to preserve agricultural diversity and rural lifestyles.

The unfortunate outbreaks of BSE or "mad cow deceases" have drawn public attention to food safety. There have also been cases where infective agents have been detected in certain one lot of food. Often these agents can only be detected in laboratory assays on samples taken from the lot, and regrettably the results may become available only after the produce has reached the market, which makes a recall difficult and one has to resort to imprecise public warnings. Knowing the origin of each food item is thus essential to ensure that it is not carrying unwanted deceases, and to enable selective recalls of infected items avoiding to waste good food as a safety precaution. It will help assuring the consumers that the food they buy is of controlled origin, and that the quality control of the shop and the public authorities extends from the farm to the table.

---

① Smart dust is a hypothetical ([ˌhaɪpəʊˈθetɪkəl]adj.假设的，假定的) system of many tiny MicroElectro-Mechanical Systems (MEMS) such as sensors, robots, or other devices, that can detect, for example, light, temperature, vibration ([vaɪˈbreɪʃən]n. 振动，颤动，摇动，摆动), magnetism or chemicals; are usually networked wirelessly; and are distributed over some area to perform tasks, usually sensing.

Should a food related disease be detected the traceability of the eaten food will enable faster detection of the origin of the infection and thus curbing its impact better and faster.

Finally, traceability may provide market feedback to the producers in a sector where the production is often planned well in advance according to wholesale dealers' prediction of the market for certain produce and the producers' flexibility is limited by long term contracts and politically decided production subsidies. The recent global food crisis highlighted that the feedback mechanisms in food market do not work as well as in other commodity markets, which makes the food availability oscillate between periods of overproduction and shortage. All the major food producers in the world could have augmented their production had they only seen the increasing demand earlier. Knowing what the market buys could stimulate the farmers to time their produce and offerings better to market demand fluctuations. The social impact of improved food supply stability can not be underestimated as hunger is a strong driving force for social unrest and uprising.

New Words

barrier	['bæriə]	n. 障碍；障碍物，栅栏，屏障
improvement	[im'pru:vmənt]	n. 改进，进步
transmission	[trænz'miʃən]	n. 播送，发射，传输，转播
metal	['metl]	n. 金属
liquid	['likwid]	n. 液体，流体
replacement	[ri'pleismənt]	n. 代替者，复位，交换，置换，移位
coexistence	[ˌkəuig'zistəns]	n. 共存
mechanism	['mekənizəm]	n. 机械装置，机构，机制
render	['rendə]	vt. 实施
		vi. 给予补偿
multiple	['mʌltipl]	adj. 多样的，多重的
		n. 倍数，若干
		v. 成倍增加
benefit	['benifit]	n. 利益，好处
		vt. 有益于，有助于
		vi. 受益
unified	['ju:nifaid]	adj. 统一的，统一标准的，一元化的
producer	[prə'dju:sə]	n. 生产者，制作者
cashier	[kæ'ʃiə]	n. (商店等的)出纳员，收款柜台
refill	[ˌri:'fil]	v. 再装满，补充，再充填
		n. 新补充物，替换物
wholesaler	['həulˌseilə]	n. 批发商
infringe	[in'frindʒ]	v. 破坏，侵犯，违反

purchase	[ˈpəːtʃəs]	vt. & n.	买，购买
dispute	[disˈpjuːt]	v.	争论，辩论，怀疑，阻止
		n.	争论，辩论，争吵
discount	[ˈdiskaunt]	n.	折扣
insurance	[inˈʃuərəns]	n.	保险，保险单，保险业，保险费
household	[ˈhaushəuld]	n.	家庭，家族
		adj.	家庭的，家族的，普通的，平常的
consult	[kənˈsʌlt]	v.	翻阅，查阅，参考，考虑
innovation	[ˌinəuˈveiʃən]	n.	改革，创新
fundamentally	[fʌndəˈmentəli]	adv.	基础地，根本地
proactive	[ˌprəuˈæktiv]	adj.	积极主动的，主动出击的
intervention	[ˌintə(ː)ˈvenʃən]	n.	干涉
optimize	[ˈɔptimaiz]	vt.	使最优化
delegate	[ˈdeligit]	n.	代表
		vt.	委派……为代表
assemble	[əˈsembl]	vt.	集合，聚集，装配
pharmaceutical	[ˌfɑːməˈsjuːtikəl]	n.	药物
		adj.	制药(学)上的
drug	[drʌg]	n.	药
biodegradable	[ˌbaiəudiˈgreidəbl]	adj.	生物所能分解的
pill	[pil]	n.	药丸
abuse	[əˈbjuːz]	n. & v.	滥用
dust	[dʌst]	n.	灰尘
incompatible	[ˌinkəmˈpætəbl]	adj.	不兼容的；不相容的；互斥的
overdose	[ˈəuvədəus]	n.	配药量过多
		vt.	配药过量，使服药过量
accident	[ˈæksidənt]	n.	意外事件，事故
perish	[ˈperiʃ]	vi.	毁灭，死亡
misuse	[ˌmisˈjuːz]	v.	误用，错用，滥用
	[misˈjuːz]	n.	误用，错用；滥用
antidote	[ˈæntidəut]	n.	解毒剂，矫正方法
dosage	[ˈdəusidʒ]	n.	剂量，配药，用量
taste	[teist]	v.	品尝，辨味
		n.	味道，味觉
traceable	[ˈtreisəbl]	adj.	可追踪的，起源于
diversity	[daiˈvəːsiti]	n.	差异，多样性
outbreak	[ˈautbreik]	n.	(战争的)爆发，(疾病的)发作
assay	[əˈsei]	n. & v.	化验
regrettably	[riˈgretəbli]	adv.	抱歉地，遗憾地，可悲地

recall	[ri'kɔːl]	vt. & n. 召回
disease	[di'ziːz]	n. 疾病，弊病
traceability	[ˌtreisə'biləti]	n. 可追溯，可描绘，可描写
feedback	['fiːdbæk]	n. 反馈，反应
prediction	[pri'dikʃən]	n. 预言，预报
flexibility	[ˌfleksə'biliti]	n. 弹性，适应性，机动性，柔性
highlight	['hailait]	vt. 突出
oscillate	['ɔsileit]	v. 振荡
overproduction	[ˌəuvəprə'dʌkʃən]	n. 生产过剩
shortage	['ʃɔːtidʒ]	n. 不足，缺乏
augmented	[ɔːg'mentid]	adj. 扩张的
stimulate	['stimjuleit]	v. 刺激，激励
stability	[stə'biliti]	n. 稳定性
underestimate	[ˌʌndər'estimeit]	vt. & n. 低估
unrest	[ʌn'rest]	n. 不安定，动荡，骚乱
uprising	['ʌpˌraiziŋ]	n. 暴动；升起

✎ Phrases

pilot project	(小规模)试验计划
shop floor	车间，工场
check out	结账；检验，合格
theft protection	防盗
conveyer belt	输送带
quality management	质量管理
supply chain	供应链
bar code	条形码
in somebody's name	以某人的名义
sort out	挑选出
capital goods	生产资料；资本财货，资本货物
marked with …	以……为标记，以……表明
on behalf of …	代表……
in transit	运送中的
smart dust	智能微尘
active substance	有效物质，有效成分
quest for	追求，探索
mad cow disease	疯牛病
public authority	政府当局
commodity market	商品市场

Abbreviations

NFC (Near Field Communication) 近场通信，近距离无线通信技术
BSE (Bovine Spongiform Encephalopathy) 牛绵状脑病，疯牛病

Exercises

I. Answer the following questions according to the text.

1. What will the first large scale application of the Internet of Things technologies be?
2. How will the retailers have item identification unified from the producer?
3. What benefits do the electronic tags offer over the bar code for the consumers?
4. What is almost every phone sold equipped with today?
5. Does the innovation in logistics normally change the industry fundamentally? What does it do?
6. What do the executable codes in the tags enable the *thing* in transit to do?
7. What is present day logistics based on?
8. What may smart biodegradable dust embedded inside pills do?
9. What happened after the outbreaks of BSE or "mad cow deceases"?
10. What did the recent global food crisis highlight?

II. Translate the following terms or phrases from English into Chinese and vice versa.

1. fundamentally
2. coexistence
3. feedback
4. 供应链
5. intervention
6. 条形码
7. incompatible
8. assemble
9. stability
10. flexibility

Text A 参考译文

物 联 网

物联网(IoT)是指可唯一标识的物体及其在类似因特网结构中的虚拟表现。物联网这一术语是 Kevin Ashton 于 1999 年首先使用的。物联网的概念通过 Auto-ID 中心及相关市场分

析出版物首先流行开来。射频识别(RFID)通常被视为物联网的先决条件。如果生活中的所有物体都带有无线标签，那么它们就可以被计算机识别和存储。但是，物体的唯一标识也可以用其他方法来实现，如条形码或二维码。

如果世界上的全部物体都配备了微小识别设备，那么地球上的日常生活将经历一场变革。公司不会缺货或者浪费产品，因为相关的各方都可以准确地了解他们所需的和消耗的产品。配备微小识别设备也易于追踪和定位遗失及失窃的产品。

1. 多重定义

目前已经出现了多种物联网的定义，且随着这些观点的实现和技术的发展，该术语也在发生变化。以下几种定义有部分重叠。

1.1 CASAGRAS

CASAGRAS 定义物联网是一个全球化的网络基础结构，它使用数据捕捉和通信功能把物理和虚拟物体链接起来。这个基础结构包括现有的和进化中的因特网以及网络发展。它将提供特殊的物体识别、传感器和连接能力，以此作为独立协作服务和应用的发展基础。因此，CASAGRAS 具有高度自治的数据捕捉、事件传输、网络连通与协同工作的特点。

1.2 SAP

SAP 认为物联网可看做是一个物理对象被无缝整合到信息网络的世界，在那里物理对象可以主动参与业务进程。SAP 提供了可通过网络来与这些"智能物体"交互、查询和改变它们的状态以及与它们相关的任何信息的服务，并考虑了服务中的安全和私密问题。

1.3 EPoSS

EpoSS 认为物联网是由物/物体构成的一种网络。在其运行的智能空间中，这些物体具有身份和虚拟品质，它们通过智能接口与用户、社会和环境连接及通信。

1.4 CERP-IoT

物联网是未来因特网组成的一部分，可以定义为动态网络基础结构，具有基于标准和协调通信协议的自配置能力。在物联网中，物理和虚拟的"物"都有标识、物理属性及虚拟品质并使用了智能接口，而且都可以无缝连接到信息网络中。物联网中的"物"可以成为业务、信息和社会进程的主动参与者，它们可以相互交互与通信，并可通过交换环境"感知"到的信息和数据与环境通信，同时自动地响应"真实/物理"事件，然后通过运行能够触发行为和建立有人或无人干预的服务程序来影响环境。这种服务接口可以促进这些"智能物体"在因特网上的交互、查询并改变它们的状态及任何相关信息，同时保证信息的安全与私密。

1.5 其他

未来的物联网将把唯一可标识的物与其在因特网中的实际表现相链接，因特网包括或连接了附带它们身份、状态、位置或其他业务的信息以及社会的或私人的相关的金融与非金融的支付信息。这不仅提供了信息也供非预定的共享者访问。所提供的精确和适当的信息可以按恰当的数量和条件、恰当的时间和地点、以恰当的代价来访问。物联网与以下概念意义不同：泛在网/普适计算、因特网协议、通信技术、嵌入式设备及其应用、人的因特网或者物的内联网/外联网，但物联网对这些都有依赖。智能虚拟表现(如所谓形象化和嵌入式、驻留云中的或集中的应用)与物理物体的结合有时也叫做"计算体"。因此,尽管计算体是由人编写的程序控制的,但它们还是被当作价值链上的自治角色,其能够感知、分析并对各种环境做出响应。计算体可起助手、顾问和决策者等作用,也可当作真正的(经济学意义的)代理,帮助改变现有的经济或组织模式。在这种情况下,形象化的意思就是人工智能和复杂系统。

2. 物的唯一可寻址能力

Auto-ID 中心的初始想法是基于 RFID 标签和电子产品代码的唯一表示标识性的。

换个角度,从语义网来看,所有的东西(不仅是电子的、智能的或嵌入 RFID 的)都可以通过现有的命名协议(如 URI)来寻址。物体本身不会交流,但可以通过其他代理(例如扮演它们属主的强力中心服务器)来联系。

使用因特网协议第六版(IPv6)的下一代因特网应用能够与所有附带人造物体的设备通信,因为 IPv6 有巨大的地址空间。因此,这个系统有能力识别多种物体。

可以在 GS1/EPCglobal EPC 信息服务规范中找到上面想法的组合。目前,这个系统已经用来识别工业中的物体,其应用范围从宇宙飞船到快速消费品和物流运输。

3. 趋势与特点

3.1 智能

环境智能和自治控制并不是物联网初始概念中的一部分,也不是因特网结构所必需的。但是,物联网的研究已经转移到整合物联网和自治控制方面了。物联网的未来可能不确定,可能是一个带有自组织或智能实体(Web 服务、SOA 成分)、虚拟物体的开放网络,它们能够协调和独立地起作用(追求自己的目标或共享目标),这取决于上下文、周边及外部环境。

嵌入智能呈现了物联网"面向人工智能"的前景,可以更清晰地定义为:利用该性能去收集和分析人与广泛存在的智能物体交互时留下的数字轨迹,以便发现人类生活、与环境交互及社会联系/行为的知识。

3.2 体系结构

物联网可能是事件驱动体系、自底向上(基于实时处理和运作环境)并考虑任何辅助等级的一个示例。因此,模型驱动和功能路径会与新的能够处理例外和不寻常的发展路径的方法共存。

3.3 复杂系统

在半开或封闭环路中,因为在自治角色中有大量的链接和交互并能够整合新的角色,所以会将物联网当作复杂系统考虑和研究。在全进程(全开环路)中,物联网可能被看做是一个混沌环境。

3.4 规模考虑

物联网可以编码 50 到 100 万亿的物体,并能跟踪这些物体的移动。

3.5 时间考虑

在这种物联网中,有数以十亿计的并行和并发事件,时间不再被用作普通和线性维度,而是取决于每个实体(物体、进程、信息系统等)。因此,该物联网将基于大量并行的IT系统(并行计算)运行。

3.6 空间考虑

在一个物联网中,一个物体的精确的地理位置(即精确的地理维度)是关键。现在,因特网主要用于管理人们处理的信息。因此,一个物体的实际情况(如其时空位置)对于追踪已经不那么重要了,因为处理这些信息的人可以确定这些信息对于行动是否重要。如此,就可以减少不必要的信息(或者决定不采取行动)。注意,物联网中的有些物体是传感器,而传感器的位置通常是非常重要的。当这些物体可以按照位置组织和连接时,GeoWeb和Digital Earth的应用就可能会很有前景。但是,也要面对这些挑战:可变的空间规模约束、需要处理的大量数据、快速检索以及相邻操作的索引。在物联网中,如果物体可以主动地采取行动,则以人为中心的协调角色就可以取消,而且在这个信息生态环境中,必须赋予我们人类不以为然的时—空环境一个重要角色。正如标准在因特网和网络中扮演的关键角色一样,地理空间标准将在物联网中扮演关键角色。

4. 框架

物联网框架可以为"物"之间的交互提供支撑而且可以涉及更复杂的结构,如分布式计算和分布式应用的发展。当下,物联网框架似乎集中在像Pachube这样的实时数据记录解决方案中:提供与许多"物"一起工作及相互交互的基准。未来可能要开发特殊的软件开发环境,以便建立在物联网中与硬件协调工作的软件。

Unit 2 Internet

Text A

Internet

1. How does the Internet Work?

Even though the Internet is still a young technology, it's hard to imagine life without it now. Every year, engineers create more devices to integrate with the Internet. This network of networks crisscrosses the globe and even extends into space. But what makes it work?(See Figure 2.1)

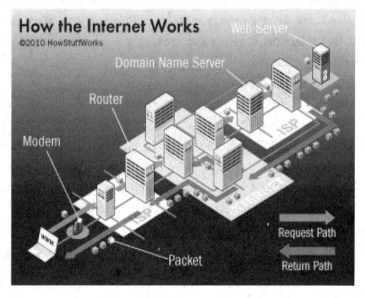

Figure 2.1 Internet architecture

To understand the Internet, it helps to look at it as a system with two main components. The first of those components is hardware. That includes everything from the cables that carry terabits of information every second to the computer sitting in front of you.

Other types of hardware that support the Internet include routers, servers, cell phone towers, satellites, radios, smartphones① and other devices. All these devices together create the network of networks. The Internet is a malleable system—it changes in little ways as elements join and leave networks around the world. Some of those elements may stay fairly static and make up the backbone of the Internet. Others are more peripheral.

These elements are connections. Some are end points—the computer, smartphone or other device you're using to read this may count as one. We call those end points clients. Machines that store the information we seek on the Internet are servers. Other elements are nodes which serve as a connecting point along a route of traffic. And then there are the transmission lines which can be physical, as in the case of cables and fiber optics.

All of this hardware wouldn't create a network without the second component of the Internet: the protocols. Protocols are sets of rules that machines follow to complete tasks. Without a common set of protocols that all machines connected to the Internet must follow, communication between devices couldn't happen. The various machines would be unable to understand one another or even send information in a meaningful way. The protocols provide both the method and a common language for machines to use to transmit data.

2. A Matter of Protocols

You've probably heard of several protocols on the Internet. For example, hypertext transfer protocol is what we use to view Web sites through a browser—that's what the http at the front of any Web address stands for. If you've ever used an FTP server, you relied on the file transfer protocol. Protocols like these and dozens more create the framework within which all devices must operate to be part of the Internet.

Two of the most important protocols are the Transmission Control Protocol (TCP)② and the Internet Protocol (IP)③. We often group the two together—in most discussions about Internet Protocols you'll see them listed as TCP/IP.

What do these protocols do? At their most basic level, these protocols establish the rules for how information passes through the Internet. Without these rules, you would need direct connections to other computers to access the information they hold. You'd also need both your computer and the target computer to understand a common language.

You've probably heard of IP addresses. These addresses follow the Internet protocol. Each

---

① A smartphone is one device that can take care of all of your handheld computing and communication needs in a single, small package. It's not so much a distinct class of products as it is a different set of standards for cell phones (蜂窝电话) to live up to.
② TCP (Transmission Control Protocol) is a set of rules (protocol) used along with the Internet Protocol (IP) to send data in the form of message units between computers over the Internet. While IP takes care of handling the actual delivery ([di'livəri] n.发送,传输) of the data, TCP takes care of keeping track of the individual units of data (called packets) that a message is divided into for efficient routing through the Internet.
③ The Internet Protocol (IP) is the method or protocol by which data is sent from one computer to another on the Internet. Each computer (known as a host ([həust]n.主机)) on the Internet has at least one IP address that uniquely identifies it from all other computers on the Internet.

device connected to the Internet has an IP address. This is how one machine can find another through the massive network.

The version of IP most of us use today is IPv4, which is based on a 32-bit address system. There's one big problem with this system: We're running out of addresses. That's why the Internet Engineering Task Force (IETF) decided back in 1991 that it was necessary to develop a new version of IP to create enough addresses to meet demand. The result was IPv6, a 128-bit address system. That has enough addresses to accommodate the rising demand for Internet access for the foreseeable future.

When you want to send a message or retrieve information from another computer, the TCP/IP protocols are what make the transmission possible. Your request goes out over the network, hitting Domain Name Servers (DNS)① along the way to find the target server. The DNS points the request in the right direction. Once the target server receives the request, it can send a response back to your computer. The data might travel a completely different path to get back to you. This flexible approach to data transfer is part of what makes the Internet such a powerful tool.

3. Packet

In order to retrieve this article, your computer has to connect with the Web server containing the article's file. We'll use that as an example of how data travels across the Internet.

First, you open your Web browser and connect to our Web site. When you do this, your computer sends an electronic request over your Internet connection to your Internet Service Provider (ISP). The ISP routes the request to a server further up the chain on the Internet. Eventually, the request will hit a Domain Name Server (DNS). This server will look for a match for the domain name you've typed in (such as www. howstuffworks.com). If it finds a match, it will direct your request to the proper server's IP address. If it doesn't find a match, it will send the request further up the chain to a server that has more information.

The request will eventually come to our Web server. Our server will respond by sending the requested file in a series of packets. Packets are parts of a file that range between 1,000 and 1,500 bytes. Packets have headers and footers that tell computers what's in the packet and how the information fits with other packets to create an entire file. Each packet travels back up the network and down to your computer. Packets don't necessarily all take the same path—they'll generally travel the path of least resistance.

That's an important feature. Because packets can travel multiple paths to get to their destination, it's possible for information to route around congested areas on the Internet. In fact,

---

① The Domain Name System (DNS) is a <u>hierarchical</u> ([ˌhaɪəˈrɑːkɪkəl]adj.分等级的) distributed naming system for computers, services, or any resource connected to the Internet or a private network. It associates various information with domain names assigned to each of the participating entities.

A Domain Name Service translates queries for domain names (which are meaningful to humans) into IP addresses for the purpose of locating computer services and devices worldwide.

as long as some connections remain, information could still travel from one section to another——though it might take longer than normal.

When the packets get to you, your device arranges them according to the rules of the protocols. It's kind of like putting together a jigsaw puzzle. The end result is that you see this article.

This holds true for other kinds of files as well. When you send an e-mail, it gets broken into packets before zooming across the Internet. Phone calls over the Internet also convert conversations into packets using the Voice over Internet Protocol (VoIP)①. We can thank network pioneers like Vinton Cerf② and Robert Kahn for these protocols——their early work helped build a system that's both scalable and robust.

That's how the Internet works in a nutshell. As you look closer at the various devices and protocols, you'll notice that the picture is far more complex than the overview we've given.

New Words

Internet	['intənet]	n. 因特网，国际互联网络
component	[kəm'pəunənt]	n. 成分
		adj. 组成的，构成的
hardware	['hɑ:dwɛə]	n. (电脑的)硬件
cable	['keibl]	n. 电缆
carry	['kæri]	vt. 携带，运送，支持，传送
		vi. 被携带，能达到
terabit	['terəbit]	n. 兆兆位(量度信息单位)
satellite	['sætəlait]	n. 人造卫星
smartphone	['smɑ:tfəun]	n. 智能电话
device	[di'vais]	n. 装置，设备
malleable	['mæliəbl]	adj. 有延展性的
element	['elimənt]	n. 要素，元素，成分，元件
static	['stætik]	adj. 静态的，静力的
backbone	['bækbəun]	n. 中枢，骨干
peripheral	[pə'rifərəl]	adj. 外围的
		n. 外围设备
connection	[kə'nekʃən]	n. 连接，关系，接线，线路

---

① Voice over IP (VoIP) commonly refers to the communication protocols, technologies, methodologies, and transmission techniques involved in the delivery of voice communications and multimedia sessions (['seʃən] n.会话) over Internet Protocol (IP) networks, such as the Internet. Other terms commonly associated with VoIP are IP telephony, Internet telephony, Voice over BroadBand (VoBB, 宽带电话), broadband telephony, and broadband phone.

② Vinton Cerf 与 Robert Kahn 一起设计了 TCP/IP 协议，被称为互联网之父。

client	[ˈklaiənt]	n.	顾客，客户，客户机
route	[ruːt]	n.	路线，路程，通道
		v.	发送
traffic	[ˈtræfik]	n.	流量，通信量
fiber	[ˈfaibə]	n.	光纤
protocol	[ˈprəutəkɔl]	n.	协议
task	[tɑːsk]	n.	任务，作业
		v.	分派任务
meaningful	[ˈmiːniŋful]	adj.	有意义的
provide	[prəˈvaid]	v.	供应，供给
transmit	[trænzˈmit]	vt.	传输，发射，传播
		vi.	发射信号，发报
data	[ˈdeitə]	n.	数据
hypertext	[ˈhaipətekst]	n.	超文本
discussion	[disˈkʌʃən]	n.	讨论
establish	[iˈstæbliʃ]	vt.	建立，设立，确定
address	[əˈdres]	n.	地址
accommodate	[əˈkɔmədeit]	vt.	供应，供给，使适应，调节
foreseeable	[fɔːˈsiːəbl]	adj.	可预知的，能预测的
retrieve	[riˈtriːv]	n.	检索；恢复，取回
transmission	[trænzˈmiʃən]	n.	发射，传送，传输，转播
response	[riˈspɔns]	n.	回答，响应，反应
flexible	[ˈfleksəbl]	adj.	柔韧性，灵活的，柔软的
packet	[ˈpækit]	n.	数据包
request	[riˈkwest]	vt. & n.	请求，要求
eventually	[iˈventʃuəli]	adv.	终于，最后
match	[mætʃ]	n.	匹配
file	[fail]	n.	文件
remain	[riˈmein]	vi.	保持，逗留，剩余
convert	[kənˈvəːt]	vt.	使转变，转换……
conversation	[ˌkɔnvəˈseiʃən]	n.	会话，交谈
scalable	[ˈskeiləbl]	adj.	可升级的

✎ Phrases

integrate with …	使与……结合，与……整合
look at … as	把……看做
end point	端点

transmission line	传输线
set of rules	规则组，规则集
one another	互相，彼此
in a meaningful way	以有意义的方式
hypertext transfer protocol	超文本传输协议
stand for	代表，表示；代替
pass through	经过，通过
run out of	用完
meet demand	满足需求
target server	目标服务器
connect with	连接，联络
a series of	一连串的，一系列的
kind of	有点儿，有几分
jigsaw puzzle	七巧板，智力拼图玩具
hold true	适用，有效
in a nutshell	简单地，简约地

Abbreviations

FTP (File Transfer Protocol)	文件传输协议
TCP (Transmission Control Protocol)	传输控制协议
IP (Internet Protocol)	因特网协议
IETF (Internet Engineering Task Force)	因特网工程工作小组
DNS (Domain Name Server)	域名服务器
ISP (Internet Service Provider)	因特网服务提供商

Exercises

I. Answer the following questions according to the text.

1. What are the two main components of Internet?

2. What are servers?

3. What are protocols?

4. What are two of the most important protocols?

5. What does the http at the front of any Web address stand for?

6. How can one machine find another through the massive network?

7. What is the version of IP most of us use today? What is one big problem with this system?

8. What are packages? How do they generally travel?

9. Why is it possible for information to route around congested areas on the Internet?

10. How do phone calls over the Internet also convert conversations into packets?

II. Translate the following terms or phrases from English into Chinese and vice versa.

1. packet
2. conversation
3. fiber
4. 超文本传输协议
5. route
6. 传输线
7. protocol
8. peripheral
9. transmit
10. device

III. Fill in the blanks with the words given below.

| meet | resources | office | technologies | secure |
| offer | relationships | remote | protected | ordering |

◆◆◆ What is an Extranet? ◆◆◆

An extranet is similar to an intranet but it is made accessible to selected external partners such as business partners, suppliers, key customers, etc, for exchanging data and applications and sharing information.

As with an intranet, an extranet can also provide ___1___ access to corporate systems for staff who spend lots of time out of the ___2___, for instance those in sales or customer support, or home workers.

Extranet users should be a well-defined group and access must be ___3___ by rigorous identification routines and security features.

Here are some reasons why businesses use extranet. Businesses of all sizes are under increasing pressure to use online ___4___, electronic order tracking and inventory management.

At the same time small businesses are keen to ___5___ the demands of larger companies in terms of working flexibly, adopting new ___6___ and enabling the exchange of business information and transactions.

Extranets ___7___ a cheap and efficient way for businesses to connect with their trading partners. It also means that your business partners and suppliers can access the information they need 24 hours a day.

The ability of the extranet to automate the trading tasks between you and your trading partners can lead to enhanced business ___8___ and help to integrate your business firmly within their supply chain.

And now let's come to the key issues to consider when using extranet.

Bear in mind that once you make parts of your business data available to trading partners, they will expect it to be available, up to date and ___9___ at all times. Availability and security are key factors in the success of an extranet.

Is your business prepared to work collaboratively? Are you willing to share production and sales information with partners?

Significant ___10___ may be required to keep the content of the extranet accurate and up to date. This might involve the use of content management software and training for staff to use it.

IV. Translate the following passage from English to Chinese.

◇◇◇ Home Page ◇◇◇

For a Web user, the home page is the first Web page that is displayed after starting a Web browser like Netscape's Navigator or Microsoft's Internet Explorer. The browser is usually preset so that the home page is the first page of the browser manufacturer. However, you can set it to open to any Web site. For example, you can specify that "http://www.yahoo.com" or "http://whatis.com" be your home page. You can also specify that there be no home page (a blank space will be displayed) in which case you choose the first page from your bookmark list or enter a Web address.

Network Device

1. Firewall

A firewall is a set of related programs located at a network gateway① server. It protects the resources of a private network from users from other networks. The term also implies the security policy that is used with the programs. An enterprise with an intranet② that allows its workers to have access to the wider Internet installs a firewall to prevent outsiders from accessing its own

---

① A gateway is a network point that acts as an <u>entrance</u> ([in'trɑ:ns]n.入口, 进入) to another network. On the Internet, a node or stopping point can be either a gateway node or a host (end-point) node. Both the computers of Internet users and the computers that serve pages to users are host nodes. The computers that control traffic within your company's network or at your local Internet Service Provider (ISP) are gateway nodes.

② An intranet is a private network that is contained within an enterprise. It may consist of many <u>interlinked</u> ([,intə(:)'liŋk] vt. 连接) local area networks and also use leased lines in the wide area network. Typically, an intranet includes connections through one or more gateway computers to the outside Internet. The main purpose of an intranet is to share company information and computing resources among employees.

private data resources and for controlling what outside resources its own users have access to.

Basically, a firewall, working closely with a router program, examines each network packet to determine whether to forward it toward its destination. A firewall also includes or works with a proxy server that makes network requests on behalf of workstation users. A firewall is often installed in a specially designated computer separate from the rest of the network so that no incoming request can get directly at private network resources.

There are a number of firewall screening methods. A simple one is to screen requests to make sure they come from acceptable (previously identified) domain name and Internet Protocol addresses. For mobile users, firewalls allow remote access in to the private network by the use of secure logon procedures and authentication certificates.

A number of companies make firewall products. Features include logging and reporting, automatic alarms at given thresholds of attack, and a graphical user interface for controlling the firewall.

Computer security borrows this term from firefighting, where it originated. In firefighting, a firewall is a barrier established to prevent the spread of fire.

2. Gateway

A gateway is a network point that acts as an entrance to another network. On the Internet, a node or stopping point can be either a gateway node or a host (end-point) node. Both the computers of Internet users and the computers that serve pages to users are host nodes. The computers that control traffic within your company's network or at your local Internet Service Provider (ISP) are gateway nodes.

In the network for an enterprise, a computer server acting as a gateway node is often also acting as a proxy server and a firewall server. A gateway is often associated with both a router, which knows where to direct a given packet of data that arrives at the gateway, and a switch, which furnishes the actual path in and out of the gateway for a given packet.

3. Router

In packet-switched networks such as the Internet, a router is a device or, in some cases, software in a computer that determines the next network point to which a packet should be forwarded toward its destination. The router is connected to at least two networks and decides which way to send each information packet based on its current understanding of the state of the networks it is connected to. A router is located at any gateway where one network meets another, including each point-of-presence on the Internet. A router is often included as part of a network switch.

A router may create or maintain a table of the available routes and their conditions and use this information along with distance and cost algorithms to determine the best route for a given packet. Typically, a packet may travel through a number of network points with routers before arriving at its destination. Routing is a function associated with the Network layer (layer 3) in the

Open Systems Interconnection (OSI) model. A layer-3 switch is a switch that can perform routing functions.

An edge router is a router that interfaces with an Asynchronous Transfer Mode (ATM)[①] network. A brouter is a network bridge combined with a router.

4. Bridge

In telecommunication networks, a bridge is a product that connects a Local Area Network (LAN) to another local area network that uses the same protocol (for example, Ethernet or Token Ring[②]). You can envision a bridge as being a device that decides whether a message from you to someone else is going to the local area network in your building or to someone on the local area network in the building across the street. A bridge examines each message on a LAN, "passes" those known to be within the same LAN, and forwards those known to be on the other interconnected LAN (or LANs).

In bridging networks, computer or node addresses have no specific relationship to location. For this reason, messages are sent out to every address on the network and accepted only by the intended destination node. Bridges learn which addresses are on which network and develop a learning table so that subsequent messages can be forwarded to the right network.

Bridging networks are generally always interconnected local area networks since broadcasting every message to all possible destinations would flood a larger network with unnecessary traffic. For this reason, router networks such as the Internet use a scheme that assigns addresses to nodes so that a message or packet can be forwarded only in one general direction rather than forwarded in all directions.

A bridge works at the data-link (physical network) level of a network, copying a data frame from one network to the next network along the communications path.

A bridge is sometimes combined with a router in a product called a brouter.

5. Hub

In general, a hub is the central part of a wheel where the spokes come together. The term is familiar to frequent fliers who travel through airport "hubs" to make connecting flights from one point to another. In data communications, a hub is a place of convergence where data arrives from one or more directions and is forwarded out in one or more other directions. A hub usually includes a switch of some kind. And a product that is called a "switch" could usually be considered a hub as well. The distinction seems to be that the hub is the place where data comes together and the switch is what determines how and where data is forwarded from the place

---

① ATM(Asynchronous Transfer Mode) is a dedicated (['dedikeitid]adj.专注的)-connection switching technology that organizes digital data into 53-byte cell units and transmits them over a physical medium using digital signal technology.

② A Token Ring network is a Local Area Network (LAN) in which all computers are connected in a ring or star topology (星型拓扑) and a bit- or token-passing scheme is used in order to prevent the collision of data between two computers that want to send messages at the same time.

where data comes together. Regarded in its switching aspects, a hub can also include a router.

• In describing network topologies, a hub topology consists of a backbone (main circuit) to which a number of outgoing lines can be attached ("dropped"), each providing one or more connection port for device to attach to. For Internet users not connected to a local area network, this is the general topology used by your access provider. Other common network topologies are the bus network and the ring network. (Either of these could possibly feed into a hub network, using a bridge.)

• As a network product, a hub may include a group of modem cards for dial-in users, a gateway card for connections to a local area network (for example, an Ethernet or a Token Ring), and a connection to a line.

6. Switch

In a telecommunications network, a switch is a device that channels incoming data from any of multiple input ports to the specific output port that will take the data toward its intended destination. In the traditional circuit-switched telephone network, one or more switches are used to set up a dedicated though temporary connection or circuit for an exchange between two or more parties. On an Ethernet LAN, a switch determines from the physical device (Media Access Control or MAC) address in each incoming message frame which output port to forward it to and out of. In a wide area packed-switched network such as the Internet, a switch determines from the IP address in each packet which output port to use for the next part of its trip to the intended destination.

In the Open Systems Interconnection (OSI①) communications model, a switch performs the layer 2 or Data-Link layer② function. That is, it simply looks at each packet or data unit and determines from a physical address (the "MAC address") which device a data unit is intended for and switches it out toward that device. However, in wide area networks such as the Internet, the destination address requires a look-up in a routing table by a device known as a router. Some newer switches also perform routing functions (layer 3 or the Network layer functions in OSI) and are sometimes called IP switches.

On larger networks, the trip from one switch point to another in the network is called a hop. The time a switch takes to figure out where to forward a data unit is called its latency. The price paid for having the flexibility that switches provide in a network is this latency. Switches are

---

① OSI is a standard description or "reference model"(参考模型) for how messages should be transmitted between any two points in a telecommunication network. Its purpose is to guide product implementors so that their products will consistently work with other products. The reference model defines seven layers of functions that take place at each end of a communication.

② Layer 2 refers to the Data Link layer of the commonly-referenced multilayered (['mʌlti,leiə]n.多层) communication model, Open Systems Interconnection (OSI). The Data Link layer is concerned with moving data across the physical links in the network. In a network, the switch is a device that redirects data messages at the layer 2 level, using the destination Media Access Control (MAC) address to determine where to direct the message.

found at the backbone and gateway levels of a network where one network connects with another and at the subnetwork level where data is being forwarded close to its destination or origin. The former are often known as *core switches* and the latter as desktop switches.

In the simplest networks, a switch is not required for messages that are sent and received within the network. For example, a local area network may be organized in a Token Ring or bus[①] arrangement in which each possible destination inspects each message and reads any message with its address.

New Words

router	['rautə]	n. 路由器
protect	[prə'tekt]	vt. 保护，关税保护，投保
enterprise	['entəpraiz]	n. 企业；公司
examine	[ig'zæmin]	vt. 检查，调查；研究，分析
screening	['skri:niŋ]	n. 筛选，屏蔽
secure	[si'kjuə]	adj. 安全的，可靠的
logon	['ləugən,'lɔg,ɔn]	v. 登录上网
logging	['lɔgiŋ]	n. 存入，联机；记录
alarm	[ə'lɑ:m]	n. 警报，警告器
		vt. 恐吓，警告
given	['givn]	adj. 约定的；特定的；指定的
threshold	['θreʃhəuld]	n. 门槛；阈值；起点，开端
term	[tə:m]	n. 术语
firefighting	['faiə,faitiŋ]	n. 消防
node	[nəud]	n. 节点
furnish	['fə:niʃ]	v. 供给
presence	['prezns]	n. 存在，到场，出席
cost	[kɔst]	n. 成本，价钱，代价
brouter	['brautə]	n. 桥式路由器
bridge	[bridʒ]	n. 网桥
Ethernet	['i:θənet]	n. 以太网
interconnect	[,intəkə'nekt]	vt. 使互相连接
subsequent	['sʌbsikwənt]	adj. 后来的，并发的
flood	[flʌd]	n. 洪水，水灾
		vt. 淹没
		vi. 被水淹，涌进

---

① In a computer or on a network, a bus is a <u>transmission path</u> (传输路径) on which signals are dropped off or picked up at every device attached to the line. Only devices addressed by the signals pay attention to them; the others discard the signals.

direction	[di'rekʃən]	n. 收件人地址
frame	[freim]	n. 帧，框架，画面
hop	[hɔp]	v. 中继段
subnetwork	[sʌb'netwə:k]	n. 子网
inspect	[in'spekt]	vt. 检查，审查

📖 Phrases

be located at	位于
gateway server	网关服务器
security policy	安全策略
protect from/against	防止……遭受……；使……免于，保护……使其不受
prevent … from	阻止；制止
have access to	有权使用
data resources	数据资源
outside resources	外部资源
router program	路由程序
separate … from	分离，分开
mobile user	移动用户
remote access	远程访问
logon procedure	登录规程
authentication certificate	证书
at the threshold of	在……的开始
graphical user interface	图形用户界面
computer security	计算机安全
packet-switched network	包交换网络
in some cases	在某些情况下
network switch	网络转接
network point	网点
best route	最佳路由
arrive at	到达
layer-3 switch	第三层交换
edge router	边式路由器
network bridge	网桥
Local Area Network (LAN)	局域网
Token Ring	令牌网
send out	发送
data-link level	数据链路层
bus network	总线网络

ring network	环形网络
dial in	拨号，拨入
input port	输入端口
output port	输出端口
circuit switched	线路交换的
physical address	物理地址
Wide Area Network (WAN)	广域网
look-up	查表
core switch	中心交换
desktop switch	桌面交换

Abbreviations

OSI (Open Systems Interconnection)	开放式系统互联参考模型
ATM (Asynchronous Transfer Mode)	异步传输模式

Exercises

I. Fill in the blanks with the information given in the text.

1. A firewall is a set of related programs located at _____. It protects the resources of a private network from users from other networks.

2. A firewall is often installed in a specially designated computer separate from the rest of the network so that _____.

3. A gateway is _____ that acts as an entrance to another network. The computers of Internet users and the computers that serve pages to users are _____. The computers that control traffic within your company's network or at your local Internet Service Provider (ISP) are _____.

4. In packet-switched networks such as the Internet, a router is a _____ or, in some cases, software in a computer that determines the next network point to which a packet should be _____ toward its destination. A router is located at _____ where one network meets another, including each point-of-presence on the Internet.

5. An edge router is a router that interfaces with _____. A brouter is a network bridge.

6. In telecommunication networks, a bridge is a product that connects one local area network to another which uses _____. It examines each message _____, "passes" those known to be within the same LAN, and forwards those known to be on the other interconnected LAN (or LANs).

7. In data communications, a hub is a place of convergence where _____ and is forwarded out in one or more other directions.

8. In a telecommunications network, a switch is a device _____ from any of multiple input ports to the specific output port that will take the data toward its intended destination.

9. In the Open Systems Interconnection (OSI) communications model, a switch performs _____.

10. On larger networks, the trip from one switch point to another in the network is called _____.

II. Translate the following terms or phrases from English into Chinese and vice versa.

1. frame _____
2. 包交换网络 _____
3. Local Area Network (LAN) _____
4. desktop switch _____
5. Wide Area Network (WAN) _____
6. hop _____
7. network bridge _____
8. subnetwork _____
9. 环形网络 _____
10. brouter _____

1. _____
2. _____
3. _____
4. _____
5. _____
6. _____
7. _____
8. _____
9. _____
10. _____

Text A 参考译文

因 特 网

1. 因特网是如何工作的？

即使因特网仍然是年轻的技术，也难以想象现在没有因特网的生活景象。每年，工程师都会设计更多与因特网结合的设备。这个网中之网已经跨越了地球甚至扩展到太空中。但是是什么使它工作的呢？网络的结构图如图2.1所示(图略)。

要理解因特网，首先应把它看做是有两个主要部分的系统。第一部分是硬件，包括从每秒可以携带兆兆位字节的电缆到你面前的计算机。

其他支持因特网的硬件包括路由器、服务器、蜂窝电话塔、卫星、无线电、智能电话和其他设备。所有这些设备共同建立了这个网中之网。因特网是一个可延展性的系统——部件的增加和减少对其影响不大。这些部件中的一些部件保持不变，是构成因特网的骨干。除此之外，还有其他的一些外围部件。

这些外围部件是可连接的。一些是终点——计算机、智能电话或者其他可以用来读的设备(也可以算作终点)，我们称这些设备为终点设备。存储在因特网中且可找到信息的机器是服务器；其他外围部件是节点，节点是通信路径上的连接点；其次是物理的传输线路，比如电缆、光纤。

如果没有第二个部分，则这些硬件不能建立网络，这第二个部分就是协议。协议是让机器完成任务所遵循的多组规则。没有所有机器都必须遵循的、联入因特网的一组公共协议，就无法实现不同设备之间的通信。因为不同机器无法相互理解，也无法用有意义的方法发送信息。协议提供了机器用来传输数据的方法和公共语言。

2. 协议之事

你也许听说过几种因特网协议。例如，超文本传输协议就是我们通过浏览器查看网站所用的协议——就是网址前面 http 所代表的协议。你用过 FTP 的服务器就依靠了文件传输协议。这些协议和更多的协议建立了一个框架，所有作为因特网一部分运作的设备都必须在这个框架之内。

两个最重要的协议是传输控制协议(TCP)和因特网协议(IP)。我们通常把这两个协议组合在一起——讨论因特网协议时常常将其列为 TCP/IP。

这些协议可以做什么？在最基础层面，这些协议建立了通过因特网传输信息的规则。没有这些规则，你需要直接连接到另外的计算机来访问它们保存的信息，也需要在你的计算机和目标计算机之间建立它们都能理解的公共语言。

你也可能听说过 IP 地址，这些地址遵循因特网协议。每个连接到因特网的设备都有一个 IP 地址。这就是一个计算机可以通过庞大网络找到另一个计算机的方法。

当今大部分人所用的 IP 版本是 IPv4，它基于 32 位地址系统。这个系统有一个大问题：地址不够用。这就是 IETF 在 1991 年决定必须开发新 IP 版本的原因，这样才有足够的地址满足需求，结果就有了 IPv6——一个 128 位地址系统。它有足够的地址供给可预见的未来范围内增长的因特网访问量。

当我们要发送一个消息或从其他计算机上检索信息时，TCP/IP 协议使传输成为可能。通过因特网发出请求，点击途中遇到的域名服务器以找到目标服务器；DNS 把请求指向正确的方向。一旦目标服务器接收到了这个请求，就把响应送回给你的计算机。这种灵活的数据传输方法使因特网成为非常有力的工具。

3. 数据包

为了检索一篇文章，你的计算机必须连接到包含这篇文章的文件的网络服务器上。我们将使用这个例子来说明数据是如何在因特网上传输的。

首先，打开你的网络浏览器并连接到我们的网站。在你做这些的时候，你的计算机通过因特网连接向你的因特网服务提供商(ISP)发送一个电子请求。然后 ISP 把这个请求发送给因特网链上更远的服务器。最后，这个请求遇到域名服务器(DNS)。该服务器寻找与你的输入相匹配的域名(如 www.howstuffworks.com)。当它找到匹配的域名后，将把你的请求指向适当的服务器 IP 地址。如果找不到匹配的域名，它将把该请求发送给链路上更远的、有更多信息的服务器。

这个请求最终到达我们的服务器。我们的服务器响应后，用一系列数据包的形式来发送请求的文件。这些数据包是文件的一部分，大小在 1000 到 1500 字节之间。数据包也可以有头有尾，用来告诉计算机数据包里有什么以及怎样与其他数据包一起建立整个文件。

每个数据包沿网络回流并下载到你的计算机。数据包不用沿着相同的路径传输——它们一般沿阻力最小的路径传输。

数据包不用沿相同路径传输是一个重要的特点，因为数据包可以沿多个路径到达目的地。信息可能会通过因特网上一些拥挤的区域，但实际上，只要某些连接还保持，信息仍然可以从一个区域传输到另一区域——虽然可能比正常情况要多花一些时间。

当数据包到达你那里后，你的设备就会根据协议的规则排列它们。这有点像七巧板拼图。最终的结果就是你看到了这篇文章。

这对其他类型的文件也适用。当你发送一个电子邮件时，邮件在快速通过因特网之前会被分成数据包。通过因特网拨打的电话也会被转换为使用 IP 语音的数据包会话。我们应该感谢像 Vinton Cerf 和 Robert Kahn 这样的网络先驱，因为是他们建立了这些协议——他们早期的工作有助于建立一个可升级和稳固的系统。

以上就是因特网如何工作的简单描述。若你进一步考虑各种设备和协议，就会发现你看到的景象比我们给出的概述要复杂得多。

Unit 3 Architecture and Technology of IoT

Text A

Architecture, Hardware, Software and Algorithms of IoT

<u>1. Architecture</u>

The Internet of Things needs an open architecture to maximize interoperability among heterogeneous systems and distributed resources including providers and consumers of information and services, whether they be human beings, software, smart objects or devices. Architecture standards should consist of well-defined abstract data models, interfaces and protocols, together with concrete bindings to neutral technologies (such as XML, web services etc.) in order to support the widest possible variety of operating systems and programming languages.

The architecture should have well-defined and granular layers, in order to foster a competitive marketplace of solutions, without locking any users into using a monolithic stack from a single solution provider. Like the internet, the IoT architecture should be designed to be resilient to disruption of the physical network and should also anticipate that many of the nodes will be mobile, may have intermittent connectivity and may use various communication protocols at different times to connect to the IoT.

IoT nodes may need to dynamically and autonomously form peer networks with other nodes, whether local or remote and this should be supported through a decentralized, distributed approach to the architecture, with support for semantic search, discovery and peer networking. Anticipating the vast volumes of data that may be generated, it is important that the architecture also include mechanisms for moving intelligence and capabilities for Internet of Things, pattern recognition, machine learning and decision-making to enable distributed and decentralized processing of the information, either close to where data is generated or remotely in the cloud. The architectural design will also need to enable the processing, routing, storage and retrieval of events and allow for disconnected operations (e.g. where network connectivity might only be intermittent). Effective caching, pre-positioning and synchronization of requests, updates and data flows need to be an integral feature of the architecture. By developing and defining the architecture in terms of open standards, we can expect increased participation from solution providers of all sizes and a competitive marketplace that benefits end users.

Issues to be addressed:

• Distributed open architecture with end to end characteristics, interoperability of heterogeneous systems, neutral access, clear layering and resilience to physical network disruption.

• Decentralized autonomic architectures based on peering of nodes.

• Cloud computing technology, event-driven architectures, disconnected operations and synchronization.

• Use of market mechanisms for increased competition and participation.

2. Hardware

The research on nano-electronics devices will be used for implementing wireless identifiable systems with the focus on miniaturization, low cost and increased functionality. Polymers electronics[①] technology will be developed and research is needed on developing cheap, non-toxic and even disposable electronics for implementing RFID tags and sensors that include logic and analogue circuits with n and p type Thin Film Transistors (TFTs), power converters, batteries, memories, sensors, active tags.

Silicon IC technology will be used for systems with increased functionality and requirements for more non volatile memory used for sensing and monitoring ambient parameters. Research is needed on ultra-low power, low voltage and low leakage designs in submicron RF CMOS technologies, on high-efficiency DC-DC power-management solutions, ultra low power, low voltage controllable non-volatile memory, integration of RF MEMS and MEMS devices. The focus will be on highly miniaturized integrated circuits that will include:

• Multi RF, adaptive and reconfigurable Front Ends;

• HF/UHF/SHF/EHF;

• Memory –EEPROM/FRAM[②]/Polymer;

• ID 128/256 bits + other type ID;

---

① Polymer (['pɔlimə]n. 聚合体) electronics, or Polytronics, as the short name, is a promising (['prɔmisiŋ]adj. 有希望的，有前途的) technology for low-cost and large-area electronic systems, based on novel organic materials with conducting and semiconducting properties, not addressable by conventional silicon technology. The new properties of this plastic materials are a combination of the electronic and optical properties (光电性能) of metals and semiconductors and the properties of polymers and small molecules.

② FRAM (Ferroelectric Random Access Memory, 铁电存储器) is at the forefront of next generation nonvolatile (['nɔn'vɔlətail]adj. 非易失性的) memory technology, and is embedded in select MSP430 microcontrollers. Embedded FRAM is enabling faster speeds, lower power, and smaller form factors.

• Multi Communication Protocols;

• Digital Processing;

• Security, including tamper-resistance countermeasures, and technology to thwart side channel attacks.

Based on this development two trends are emerging for wireless identifiable devices for IoT applications:

• Increasing use of embedded intelligence;

• Networking of embedded intelligence.

IoT will create new services and new business opportunities for system providers to service the communication demands of potentially tens of billions of devices. Three main trends are seen today:

• Ultra low cost tags with very limited features. The information is centralized on data servers managed by service operators. Value resides in the data management.

• Low cost tags with enhanced features such as extra memory and sensing capabilities. The information is distributed both on centralized data servers and tags. Efficient network infrastructure. Value resides in communication and data management, including processing of data into actionable information.

• Smart fixed/mobile tags and embedded systems. More functions into the tag bringing local services. Smart systems (sensing/monitoring/actuating) on tags. The information is centralized on the data tag itself. Value resides in the communication management to ensure security and effective synchronization to the network. Smart devices enhanced with inter-device communication will result in smart systems with much higher degrees of intelligence and autonomy. This will enable the more rapid deployment of smart systems for IoT applications and creation of new services.

3. Software and Algorithms

One of the most promising micro operating systems for constrained devices is Contiki[①]. It provides a full IP stack (both IPv4 and IPv6), supports a local flash file system and features a large development community and a comprehensive set of development tools.

One of challenges in building IoT applications lies in the lack of a common software fabric underlying how the software in the different environments can be combined to function into a composite system and how to build a coherent application out of a large collection of unrelated software modules. Research and development is focusing on service oriented computing for developing distributed and federated applications to support interoperable machine to machine and "thing" to "thing" interaction over a network. This is based on the Internet protocols, and on

---

① Contiki is a small, open source (开源), highly portable multitasking computer operating system (多任务计算机操作系统) developed for use on a number of memory-constrained networked systems ranging from 8-bit computers to embedded systems on microcontrollers, including sensor network motes ([məuts]n. 尘埃, 微粒).

top of that, defines new protocols to describe and address the service instance. Service oriented computing loosely organizes the Web services and makes it a virtual network.

Issues to be addressed:

• Open middleware platforms;

• Energy efficient micro operating systems;

• Distributed self adaptive software for self optimization, self configuration, self healing (e.g. autonomic);

• Lightweight and open middleware based on interacting components/modules abstracting resource and network functions;

• Bio-inspired algorithms (e.g. self organization) and game theory (to overcome the risks of tragedy of commons and reaction to malicious nodes);

• Self management techniques to overcome increasing complexities;

• Password distribution mechanisms for increased security and privacy;

• Energy-aware operating systems and implementations.

New Words

distributed	[dis'tribju:tid]	adj. 分布式的
well-defined	['weldifaind]	adj. 定义明确的，明确的
abstract	['æbstrækt]	n. 摘要，概要，抽象
		adj. 抽象的
concrete	['kɔnkri:t]	adj. 具体的，有形的
binding	['baindiŋ]	n. 绑定，捆绑
neutral	['nju:trəl]	adj. 中立的
granular	['grænjulə]	adj. 由小粒而成的，粒状的
layer	['leiə]	n. 层
foster	['fɔstə]	vt. 培养，鼓励
monolithic	[,mɔnə'liθik]	n. 单片电路，单块集成电路
stack	[stæk]	n. 堆，一堆，堆栈
		v. 堆叠
resilient	[ri'ziliənt]	adj. 能复原的，有弹性的，有弹力的
disruption	[dis'rʌpʃən]	n. 中断，分裂，瓦解，破坏
anticipate	[æn'tisipeit]	v. 预订，预见，可以预料
connect	[kə'nekt]	v. 连接，联合，关联
decentralize	[di:'sentrəlaiz]	n. 分散
search	[sə:tʃ]	n. & v. 搜索，搜寻
routing	['ru:tiŋ]	n. 行程安排，邮件路由
retrieval	[ri'tri:vəl]	n. 取回，恢复，修补，重获

disconnected	[ˌdiskə'nektid]	adj. 分离的，离散的，不连贯的
synchronization	[ˌsiŋkrənai'zeiʃən]	n. 同步
update	[ʌp'deit]	v. & n. 更新
characteristic	[ˌkæriktə'ristik]	n. 特性，特征
terminal	['tə:minl]	n. 终端
nano-electronics	['nænəu-ilek'trɔniks]	n. 纳电子学
miniaturization	[ˌminiətʃərai'zeiʃən]	n. 小型化
polymer	['pɔlimə]	n. 聚合体
non-toxic	[nʌn'tɔksik]	adj. 无毒的
disposable	[dis'pəuzəbl]	adj. 可任意使用的
memory	['meməri]	n. 存储器，内存
monitor	['mɔnitə]	vt. 监控
		n. 监视器，监控器
leakage	['li:kidʒ]	n. 漏，泄漏，渗漏
submicron	['sʌb'maikrɔn]	adj. 亚微细粒的，亚微型的
controllable	[kən'trəuləbl]	adj. 可管理的，可操纵的，可控制的
configurable	[kən'figərəbl]	adj. 结构的，可配置的
tamper	['tæmpə]	v. 篡改
resistance	[ri'zistəns]	n. 反抗，抵抗，阻力，电阻，阻抗
countermeasure	['kauntəˌmeʒə]	n. 对策，反措施
opportunity	[ˌɔpə'tju:niti]	n. 机会，时机
coherent	[kəu'hiərənt]	adj. 一致的，连贯的
module	['mɔdju:l]	n. 模块
federate	['fedərit]	adj. 同盟的，联合的
instance	['instəns]	n. 实例，建议，要求，情况，场合
adaptive	[ə'dæptiv]	adj. 适应的
healing	['hi:liŋ]	n. 康复，复原
		adj. 有治疗功用的
autonomic	[ˌɔ:təu'nɔmik]	adj. 自治的，自律的
bio-inspired	['baiəu-in'spaiəd]	adj. 仿生的
tragedy	['trædʒidi]	n. 悲剧，惨案，悲惨，灾难
reaction	[ri(:)'ækʃən]	n. 反应
malicious	[mə'liʃəs]	adj. 怀恶意的，恶毒的
password	['pɑ:swə:d]	n. 密码，口令

Phrases

consist of …	由……组成
in order to	为了……

operating system	操作系统
programming language	程序设计语言
communication protocol	通信协议
peer network	对等网
pattern recognition	模式识别
machine learning	机器学习
decentralized processing	分散处理，分布处理
in terms of …	根据……，按照……，用……的话，在……方面
end to end	端对端
cloud computing technology	云计算技术
market mechanism	市场机制，市场调节职能，市场法则
analogue circuit	模拟电路
Thin Film Transistors (TFTs)	薄膜晶体管
power converter	电力变换器，整流器
active tag	有源标签，主动标签
volatile memory	非永久性存储器，易失存储器
digital processing	数字处理，数字加工
sidechannel attack	侧信道攻击
development trend	发展趋势
actionable information	全面、精确的信息
"thing" to "thing"	"物"到"物"
game theory	博弈论，对策论

Abbreviations

XML (eXtensible Markup Language)	可扩展标识语言
IC (Integrate Circuit)	集成电路
RF (Radio Frequency)	射频，无线电频率
CMOS (Complementary Metal Oxide Semiconductor)	互补金属氧化物半导体
DC (Direct Current)	直流电
MEMS (Micro-ElectroMechanical Systems)	微型机电系统(微机电)
HF (High Frequency)	高频
UHF (Ultra High Frequency)	超高频
SHF (Super High Frequency)	特高频
EHF (Extremely High Frequency)	极高频
EEPROM (Electrically Erasable Programmable Read-Only Memory)	电可擦除只读存储器
FRAM (Ferroelectric RAM)	铁电存储器

Exercises

I. Answer the following questions according to the text.

1. What does the Internet of Things need to maximize interoperability among heterogeneous systems and distributed resources?
2. What should the IoT architecture be designed?
3. What are the issues the open architecture should address?
4. What will the research on nano-electronics devices be used for?
5. What will silicon IC technology be used for?
6. What are the two trends emerging for wireless identifiable devices for IoT applications?
7. Where is the information centralized on for ultra low cost tags with very limited features? Where does value reside in?
8. For smart systems (sensing/monitoring/actuating) on tags, where is the information? Where does value reside in?
9. What is one of the most promising micro operating systems for constrained devices?
10. Where does one of challenges in building IoT applications lie in?

II. Translate the following terms or phrases from English into Chinese and vice versa.

1. Thin Film Transistors (TFTs)
2. distributed
3. module
4. communication protocols
5. stack
6. volatile memory
7. binding
8. adaptive
9. 云计算技术
10. 有源标签，主动标签

III. Fill in the blanks with the words given below.

| frequencies | storage | circuits | passive | manufactured |
| reader | range | tags | overwritten | electromagnetic |

✧✧✧ Active, Semi-passive and Passive RFID Tags ✧✧✧

Active, semi-passive and passive RFID tags are making RFID technology more accessible and prominent in our world. These ___1___ are less expensive to produce, and they can be made small enough to fit on almost any product.

Active and semi-passive RFID tags use internal batteries to power their ___2___. An active tag also uses its battery to broadcast radio waves to a reader, whereas a semi-passive tag relies on the ___3___ to supply its power for broadcasting. Because these tags contain more hardware than passive RFID tags, they are more expensive. Active and semi-passive tags are reserved for costly items that are read over greater distances—they broadcast high ___4___ from 850 to 950 MHz that can be read 100 feet (30.5 meters) or more away. If it is necessary to read the tags from even farther away, additional batteries can boost a tag's range to over 300 feet (100 meters).

Like other wireless devices, RFID tags broadcast over a portion of the ___5___ spectrum. The exact frequency is variable and can be chosen to avoid interference with other electronics or among RFID tags and readers in the form of tag interference or reader interference. RFID systems can use a cellular system called Time Division Multiple Access (TDMA) to make sure the wireless communication is handled properly.

Passive RFID tags rely entirely on the reader as their power source. These tags are read up to 20 feet (six meters) away, and they have lower production costs, meaning that they can be applied to less expensive merchandise. These tags are ___6___ to be disposable, along with the disposable consumer goods on which they are placed. Whereas a railway car would have an active RFID tag, a bottle of shampoo would have a ___7___ tag.

Another factor that influences the cost of RFID tags is data storage. There are three ___8___ types: read-write, read-only and WORM (Write Once, Read Many). A read-write tag's data can be added to or overwritten. Read-only tags cannot be added to or overwritten—they contain only the data that is stored in them when they were made. WORM tags can have additional data (like another serial number) added once, but they cannot be ___9___.

Most passive RFID tags cost between seven and 20 cents U.S. each. Active and semi-passive tags are more expensive, and RFID manufacturers typically do not quote prices for these tags without first determining their ___10___, storage type and quantity. The RFID industry's goal is to get the cost of a passive RFID tag down to five cents each once more merchandisers adopt it.

IV. Translate the following passage from English to Chinese.

DNS (Domain Name Server): DNS is the acronym for Domain Name Service, which are the machines responsible for maintaining lists that translate Internet names to numbers and vice versa. DNS allows you to reference domain names instead of their actual IP address for easier recollection.

ESN (Electronic Serial Number): The unique serial number of a cellular device that identifies it to the CDMA system for the purpose of placing and receiving calls.

MMS (Multimedia Messaging Service): Similar to SMS, but in addition to plain text. MMS messages may include multimedia elements such as pictures, video and audio. These multimedia elements are included in the message, not as attachments as with email.

Radio-Frequency fingerprinting: An electronic process that identifies each individual wireless handset by examining its unique radio transmission characteristics. Fingerprinting is used to reduce fraud since the illegal device can not duplicate the legal device's radio-frequency fingerprint.

VPN (Virtual Private Network): A VPN utilizes the public telecommunications networks to conduct private data communications. Most VPN implementations use the Internet as the public infrastructure and a variety of specialized protocols to support private communications through the Internet.

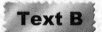

Technology of IoT

1. Identification Technology

The function of identification is to map a unique identifier or UID (globally unique or unique within a particular scope), to an entity so as to make it without ambiguity identifiable and retrievable. UIDs may be built as a single quantity or out of a collection of attributes such that the combination of their values is unique. In the vision of the Internet of Things, things have a digital identity (described by unique identifiers), are identified with a digital name and the relationships among things can be specified in the digital domain.

A unique identifier for an object can translate to a single permanent assigned name for the life of an object. However, IoT will face the need to accommodate multiple identifiers per objects, as well as changes to those identifiers. For example, many objects will have a unique identifier assigned by their manufacturer. Some may also have network addresses (such as IPv6[①] addresses), as well as temporary local identifiers within transient ad-hoc clusters of objects. Objects may also have sensors and actuators physically attached to them, with each of these sensors and actuators also being individually addressable; their identifiers may be constructed as extensions of the ID of the object or perhaps associated with the object's identifier via a lookup in a registry. Many objects may be composite objects or products that consist of replaceable parts that are exchanged during the usage phase or lifetime of the object. These parts may also have their own unique identifiers and it is important that the information models for the IoT allow changes of identifier, changes of configuration and associations between identifiers to be recorded and queried, both in terms of keeping track of changes to parent-child relationships as well as old-new relationships (e.g. where a new part is installed to replace an old part that is worn or faulty). Further examples of associations between identifiers include the breakdown of large quantities of bulk product (e.g. a specific batch of food product) into a number of individual products or packages for retail purposes, repackaging and re-labelling of products, aggregation of ingredients, components and parts to form composite products and assemblies or kits, such as medical kits.

---

① IPv6 (Internet Protocol version 6) is a set of specifications from the Internet Engineering Task Force (IETF) that's essentially an upgrade of IP version 4 (IPv4). The basics of IPv6 are similar to those of IPv4—devices can use IPv6 as source and destination addresses to pass packets over a network, and tools like ping work for network testing as they do in IPv4, with some slight variations ([ˌvɛəriˈeiʃən]n. 变更，变化).

Combinations of things will create "family tree" identification schemes where parts and components that are incorporated within composite/complex products such as computers, vehicles, and buildings have many different components, each with their own unique ID and life history. This is also referred to as a serialised Bill of Materials. This is necessary in order to track sets of different objects (e.g. parents or children of the original object) and the framework for expressing data sharing rules needs to be able to support this.

By assigning each thing participating in the Internet of Things a unique identity or potentially several unique identities, it is possible to refer to each thing as an individual, each having its own characteristics, life history and information trail, its own flow pattern through the real world and its own sequence of interactions with other things. It is important that such unique identifiers for things can be globally unique and can have significant consistency and longevity (ideally for the life of the thing), independent of the current location of the thing or the current network connectivity available to the thing, in order that it is possible to gather information about a thing even when that information is collected and owned by a number of different entities and fragmented across a large number of databases and information systems.

Many things can be considered to be (at least at the time of their creation) near-identical replicas of each other, perhaps belonging to the same product type and sharing a number of properties common to all instances within the same class of things. Often, a request or order for a particular thing might not always specify the exact unique ID that must be retrieved; instead the request can be satisfied by any thing that is a member of a particular class. It is therefore important that the Internet of Things support unique identifiers in a way that it is also possible to refer to a particular class of things as well as individual things within that class, in order to be able to retrieve or refer to class-level information and services provided for the class of things as well as serial-level information and services provided for each individual thing.

It is also important that citizens, companies and other organisations can construct unique identifiers for things as easily, affordably and autonomously as they can create unique identifiers for web pages and other internet resources, while ensuring that no two entities can claim to be the authoritative creator of the same unique ID. In the existing Internet, this is typically achieved through hierarchical identifier structures, in which each tier of the hierarchy is only responsible for ensuring uniqueness among the members of the tier below.

Familiar examples of such hierarchically structured identifiers include telephone numbers, URIs, Internet hostnames and sub domains, handles, digital object identifiers etc. It would be important to accommodate more than a single hierarchical name space; perhaps some classes of "things" would have their own name space, such as the World Wide Web using the class "IN" whose name space is managed by ICANN. Other ways that a name space can be described would be as a dominion or a realm.

However, there can be good reasons why the Internet of Things should also support "opaque" identifiers and pseudonyms, in which the internal structure of hierarchy is not readily apparent; this is particularly important when unauthorised parties are able to read the class information (e.g.

product type or object type) and could jeopardise the privacy of a citizen or the safety and security of supply chains, subjecting them to discriminatory treatment or targeted attack, on the basis of what the identifier reveals about the things which are being worn, carried or transported. There could be an opaque identifier name space that is not part of the hierarchical name space structure and reveals absolutely no information about the object that it is identifying. For example, this could have applications in uniquely identifying the medication that a patient is carrying, especially when using wireless identification technologies that lack adequate privacy measures.

We recognise that many industry sectors have already begun assigning unique identifiers to objects and that significant investment has been made in information systems and collection of information about various kinds of things, using those existing unique identifiers as keys to lookup and retrieve that information. Such established UIDs are difficult to displace and it is therefore critical for successful deployment that IoT technology can support such existing UIDs, using mapping processes where necessary.

Furthermore, as indicated in ISO 15459, multiple established name issuing authorities exist and it is important that the Internet of Things recognises their legitimate but nonexclusive involvement in the construction of unique identifiers for things and in helping to manage delegation of uniqueness of the identifiers created by their members, each of whom is thereby granted the autonomy to create unique identifiers within their own name space; it should also be possible for anyone to use Uniform Resource Identifiers (URI)[①] as unique identifiers for things.

It is important to understand that identifiers can refer to names and addresses, but since there can be multiple addresses of information and services related to an individual thing, it is probably more helpful to ensure that each thing is given a unique name and to use lookup mechanisms and referral services to obtain addresses of information and services, including those provided authoritatively by the thing's creator and those contributed by others who have interacted with the thing at some time in its life. In the case of the existence of multiple identifiers for a single object due to different reasons a scheme for ID data translation and dynamic compatibility/interoperability check is necessary.

Furthermore, it is important that identifiers are not constrained by current choices of technology for storing and communicating unique identifiers or their current limitations, since we should expect that the data carrier technology will evolve over time and current limitations (such as those on memory capacity available for identifiers) will become more relaxed.

Today various unique identifier schemes exist and interoperability is required between applications using different schemes when those applications are operated in the Future Internet environment.

The traffic in the Internet of Things networks for queries about unique identifiers will be

---

① In computing, a Uniform Resource Identifier (URI) is a string of characters (字符串) used to identify a name or a resource. Such identification enables interaction with representations of the resource over a network (typically the World Wide Web) using specific protocols. Schemes specifying a concrete syntax and associated protocols define each URI.

many times higher than that for DNS[①] queries in the current Internet.

In this context the Internet of Things deployment will require the development of new technologies that need to address the global ID schemes, identity management, identity encoding/encryption, authentication and repository management using identification and addressing schemes and the creation of global directory lookup services and discovery services for Internet of Things applications with various unique identifier schemes.

2. Communication Technology

The applications of Internet of Things form an extensive design space with many dimensions that include:

• Deployment—onetime, incremental or random.

• Mobility—occasional or continuous performed by either selected or all "things" in the selected environment.

• Cost, size, resources, and energy—very resource limited to unlimited.

• Heterogeneity—a single type of "thing" or diverse sets of differing properties and hierarchies.

• Communication modality—Electromagnetic communication—Radio Frequency, optical, acoustic, inductive and capacitive coupled communication have been used.

• Infrastructure — different applications exclude, allow or require the use of fixed infrastructure.

• Network topology—single hop, star, multihop, mesh and/or multitier.

• Coverage—sparse, dense or redundant.

• Connectivity—continuous, occasional or sporadic.

• Network size—ranging from tens of nodes to thousands.

• Lifetime—few hours, several months to many years.

• Other quality of service requirements — real time constraints, tamper resistance, unobtrusiveness

An extensive design space complicates IoT application development in various ways. One could argue that designing for the most restrictive point in the design space, e.g. minimum "thing" capabilities, highly mobile, etc. might be a solution. However, often there is no such global "minimum" and it will be desirable to exploit the characteristics of the various points in the design space. This implies that no single hardware and software platform will be sufficient to support the whole design space and heterogeneous systems will be used.

Issues to be addressed:

• Internet of Things energy efficient communications;

• Multi frequency radio front ends and protocols;

---

① The Domain Name System (DNS) is the way that Internet domain names are located and translated into Internet Protocol addresses. A domain name is a meaningful and easy-to-remember "handle" for an Internet address.

- Communication spectrum and frequency allocation;
- Software Defined Radios (SDRs)[①];
- Cognitive Radios (CRs)[②];
- Energy efficient wireless sensor networks with inter protocol communication capabilities.

3. Network Technology

The IoT deployment requires developments in network technology which is essential for implementing the vision reaching out to objects in the physical world and to bring them into the Internet. RFID, short-range wireless technologies and sensor networks are enabling this, while for example IPv6, with its expanded address space, allow that all things can be connected, and can be tracked.

In the IoT security, scalability, and cross platform compatibility between diverse networked systems will be essential. In this context the network technologies has to offer solutions that reduced costs that can offer the viability of connecting almost anything to the network, and this ubiquity of access will change the way information is processed. IP provides today end to end communication between devices, without intermediate protocol translation gateways.

Protocol gateways are inherently complex to design, manage, and deploy and with the end to end architecture of IP, there are no protocol translation gateways involved.

New scalable architectures designed specifically for the ubiquitous sensor networks communications will allow for networks of billions of devices. Improvements in techniques for secure and reliable wireless communication protocols will enable mission-critical applications for ubiquitous sensor networks based on wireless identifiable devices.

Issues to be addressed:
- Network technologies (fixed, wireless, mobile etc.);
- Ad-hoc networks.

New Words

identification	[aɪˌdentɪfɪˈkeɪʃən]	n. 辨认，鉴定，证明
entity	[ˈentɪti]	n. 实体
ambiguity	[ˌæmbɪˈgjuːɪti]	n. 含糊，不明确
retrievable	[rɪˈtriːvəbl]	adj. 可获取的

---

① A Software Defined Radio system, or SDR, is a radio communication system where components that have been typically implemented in hardware (e.g. mixers, filters ([ˈfɪltə]n. 滤波器), amplifiers ([ˈæmplɪˌfaɪə]n. 放大器), modulators/demodulators, detectors, etc.) are instead implemented by means of software on a personal computer or embedded computing devices.

② Cognitive Radio (CR) is a form of wireless communication in which a transceiver can intelligently detect which communication channels (信道) are in use and which are not, and instantly move into vacant channels while avoiding occupied ones. This optimizes the use of available Radio Frequency (RF) spectrum while minimizing interference to other users.

identifier	[aiˈdentifaiə]	n. 标识符
permanent	[ˈpəːmənənt]	adj. 永久的，持久的
assign	[əˈsain]	vt. 分配，指派
actuator	[ˈæktjueitə]	n. 驱动器，执行器
construct	[kənˈstrʌkt]	vt. 建造，构造，创立
registry	[ˈredʒistri]	n. 注册，登记
composite	[ˈkɔmpəzit]	adj. 合成的，复合的
		n. 合成物
replaceable	[riˈpleisəbl]	adj. 可代替的
configuration	[kənˌfigjuˈreiʃən]	n. 构造，结构，配置
repackage	[riˈpækidʒ]	vt. 重新包装
sequence	[ˈsiːkwəns]	n. 次序，顺序，序列
consistency	[kənˈsistənsi]	n. 一致性，连贯性
longevity	[lɔnˈdʒeviti]	n. 寿命
fragment	[ˈfrægmənt]	n. 碎片，断片，片段
dominion	[dəˈminjən]	n. 域
realm	[relm]	n. 领域
opaque	[əuˈpeik]	n. 不透明物
		adj. 不透明的
pseudonym	[ˈ(p)sjuːdənim]	n. 假名
jeopardise	[ˈdʒepədaiz]	v. 使受危险，危及
discriminatory	[diˈskriminətəri]	adj. 有差别的
treatment	[ˈtriːtmənt]	n. 待遇，对待，处理
transport	[trænsˈpɔːt]	vt. 传送，运输
displace	[disˈpleis]	vt. 取代，置换
nonexclusive	[ˈnɔniksˈkluːsiv]	adj. 非独家的
delegation	[ˌdeliˈgeiʃən]	n. 授权，委托
interoperability	[ˈintərˌɔpərəˈbiləti]	n. 互用性
random	[ˈrændəm]	n. 随意，任意
		adj. 任意的，随便的，随即的
occasional	[əˈkeiʒnəl]	adj. 偶然的，非经常的，特殊场合的，临时的
unlimited	[ʌnˈlimitid]	adj. 无限的，无约束的
modality	[məuˈdæliti]	n. 形式，形态，特征
inductive	[inˈdʌktiv]	adj. 电感的，感应的
capacitive	[kəˈpæsitiv]	adj. 电容性的
multitier	[ˈmʌltitiə]	n. 多层，多列
sparse	[spɑːs]	adj. 稀疏的
dense	[dens]	adj. 密集的
sporadic	[spəˈrædik]	adj. 零星的，孤立的

unobtrusiveness	[ˌʌnəbˈtruːsivnis]	adj. 不突出的，不引人注意的
complicate	[ˈkɔmplikeit]	v. (使)变复杂
sufficient	[səˈfiʃənt]	adj. 充分的，足够的
scalability	[ˌskeiləˈbiliti]	n. 可量测性

Phrases

digital identity	数字身份
digital domain	数字域
network address	网络地址
bulk product	散货
family tree	系谱，系谱图，族谱，族谱图
bill of materials	材料单
participate in	参加，参与，分享
in a way	在某种程度上，稍稍
hierarchical structure	层次结构，分级结构
name space	名空间
in the construction of	建筑，建造
interact with …	与……相互作用，与……相互影响；与……相互配合
single hop	单一跳跃
tamper resistance	抗干扰
software defined radio	软件定义无线电
cognitive radio	认知无线电
short-range wireless technology	短距离无线技术
cross platform compatibility	跨平台兼容性
end to end communication	端对端通信

Abbreviations

UID (User Identifier)	用户名
ID (IDentification, IDentity)	身份
URI (Uniform Resource Identifier)	统一资源标识符
ICANN (Internet Corporation for Assigned Names and Numbers)	互联网名称与数字地址分配机构

Exercises

I. Answer the following questions according to the text.

1. What is the function of identification?
2. What can a unique identifier for an object do?

3. What will combinations of things create?

4. What possibly happens by assigning each thing participating in the Internet of Things a unique identity or potentially several unique identities?

5. In order to be able to retrieve or refer to class-level information and services provided for the class of things as well as serial-level information and services provided for each individual thing, what is important?

6. What do we recognise that many industry sectors have already done?

7. Why is it probably more helpful to ensure that each thing is given a unique name and to use lookup mechanisms and referral services to obtain addresses of information and services?

8. What will the traffic in the Internet of Things networks for queries about unique identifiers be?

9. What will be essential in the IoT security, scalability, and cross platform compatibility between diverse networked systems?

10. What is said about protocol gateways? Are there any protocol translation gateways involved with the end to end architecture of IP?

II. Translate the following terms or phrases from English into Chinese and vice versa.

1. scalability
2. inductive
3. 数字身份
4. capacitive
5. identifier
6. family tree
7. 跨平台兼容性
8. tamper resistance
9. assign
10. short-range wireless technology

Text A 参考译文

物联网的体系、硬件、软件及算法

1. 体系

物联网需要开放的体系来实现不同系统和分布资源之间最大的协同性，这些资源包括信息和服务的提供者及客户，他们可以是人、软件、智能物体和设备。体系标准应该由以下几部分组成：定义好的抽象数据模型、接口和协议以及为了尽可能广泛地支持操作系统和编程语言所绑定的具体中立技术(如 XML、网络服务等)。

为了鼓励解决方案的市场竞争，体系结构应该预先定义并分成粒层，而不要把任何用户与任一个单个的解决方案提供者捆定。像互联网一样，物联网体系结构也应该具有灵活性以便适应物理网络的分裂，且应能预计许多节点的移动，能断断续续地进行连接，可以在不同的时间使用不同的通信协议来连接物联网。

物联网节点也可能需要动态地和自治地与其他节点组成对等网(无论这些节点是本地的或远程的)，这将通过分散的、分布式的方法来实现，这些方法支持语义检索、发现和对等网。该体系预计可能产生的大量数据，包括移动智能机制和物联网能力、模式识别、机器学习和决策支持以便能够实现分布式和分散的信息处理，也包括附近产生的数据和云中的远程数据，这是十分重要的。体系设计也需要能够实现事件的处理、路由选择、存储和检索并允许不连续地运行(如网络的间歇性连接)。有效的高速缓存、预定位和请求同步、更新和数据流也需要整合到体系的功能中。通过按照开放标准开发和定义体系，我们期望更多的人来参与，以提供各种规模的解决方案，并建立一个有利于最终用户的竞争市场。

建立体系需要解决的问题如下：
· 建立分布式开放体系，其应具有端对端特征，且具有协同异形系统、独立访问、清晰分层和恢复断续的物理网络的灵活性。
· 建立基于对等节点的分散自治体系。
· 建立云计算技术、事件驱动体系，以便不连续地运行和同步。
· 应使用市场机制增加竞争和参与。

2. 硬件

纳米电子设备的研究将实现无线识别系统，这些系统注重小型化、低成本并增强功能性。人们会研发聚合电子技术，并且需要开发廉价、无毒甚至一次性使用的电子设备，这些设备将实现 RFID 标签和传感器，其中包括带有 n 型和 p 型薄膜的晶体管(TFTs)、整流器、电池、存储器、传感器及主动标签的逻辑和模拟电路。

硅集成电路技术会用于带有强功能和需求的系统，这些系统更多地用于感知和监控环境参数的非易失存储器。也需要研究超低能耗、低电压和低泄露的设计方案，该设计将使用亚微型的 RF CMOS 技术、基于高效 DC-DC 能源管理方案、超低能耗、低电压可控制非易失存储器、整合 RF MEMS 与 MEMS 设备。技术焦点在于高度小型化的集成电路，包括：
· 多RF、适应的和可重构的前端；
· HF/UHF/SHF/EHF；
· 存储器——EEPROM/FRAM/Polymer；
· ID 128/256 位 + 其他类型的 ID；
· 多通信协议；
· 数字处理；
· 安全性，包括抗干扰策略以及阻止侧信道攻击的技术。

基于这个研发，出现了以下两个用于物联网的无线识别设备的新趋势：
· 嵌入式智能的广泛使用；
· 嵌入式智能的网络化。

物联网为系统提供者建立了新的服务和新的商业机会，以便为数以十亿计的设备提供通信。目前可以看到以下三个趋势：

• 功能十分有限的超低成本标签。信息集中在由服务运营者管理的数据服务器上，重点在于数据管理。

• 带有增强功能(如外部存储和感知能力)的低成本标签。信息分布在中心数据服务器和标签上，具有高效的网络基础结构。重点在于通信和数据管理，包括对可执行数据的处理。

• 智能固定/移动标签和嵌入系统。更多的功能将融入本地服务的标签。智能系统(感知/监控/行动)装在标签上，信息集中于数据标签本身。重点在于确保安全和与网络高效同步的通信管理。带有内置设备通信的增强型智能设备将产生具有更高度智能和自治的智能系统。这能够实现智能系统的更快速开发，可用于物联网应用和建立新的服务。

3. 软件和算法

用于约束设备的最有前景的微操作系统之一是 Contiki。它提供全 IP 堆栈(包括 IPv4 和 IPv6)、支持本地闪存文件系统并具有大开发社区及开发工具综合性的特点。

创建物联网应用程序所面临的挑战之一是缺乏一个公共软件基本结构。如果知道如何组合不同环境中的软件，就可以在复合的系统中开发以及如何在大量无关的软件模块中建立一个条理分明的应用。目前的开发正着重于面向服务的运算，以开发分布式和联合应用，这样就可以支持可互操作的机器对机器和"物"对"物"在网络上交互。这基于因特网协议，其中最重要的是定义新协议来描述和解决服务事宜。面向服务的计算可随意组织网络服务并使其成为一个虚拟网络。

组网需要的技术支持如下：

• 开放的中间件平台；
• 很有效的微操作系统；
• 用于自优化、自配置、自康复(如自律)的分布式自适应软件；
• 基于交互式的抽象资源和网络功能的组件/模块的轻量级和开放中间件；
• 源于生物的算法(如自组织)和博弈论(以便克服普通风险和反对恶意节点)；
• 自管理技术，以便克服日益增加的复杂性；
• 增加安全和隐私性的密码分布机制；
• 能量感知操作系统及其执行。

Unit 4　Ubiquitous Network and VPNs

Text A

How Ubiquitous Networking Will Work?

1. Introduction to How Ubiquitous Networking Will Work

Mobile computing devices have changed the way we look at computing. Laptops and Personal Digital Assistants (PDAs)① (See Figure 4.1) have unchained us from our desktop computers. A group of researchers at AT&T② Laboratories Cambridge are preparing to put a new spin on mobile computing. In addition to taking the hardware with you, they are designing a ubiquitous networking system that allows your program applications to follow you wherever you go.

Inside the bat ultrasonic transmitter shows two-copper coil antennae, a radio transmitter module, the battery and two ultrasonic transmitters.

Figure 4.1　PDA's model

---

① PDAs, also called handhelds or palmtops (['pɑːmtɔp]n.掌上型电脑), have definitely evolved over the years. Not only can they manage your personal information, such as contacts, appointments, and to-do lists, today's devices can also connect to the Internet, act as Global Positioning System (GPS, 全球定位系统) devices, and run multimedia software. What's more, manufacturers have combined PDAs with cell phones, multimedia players and other electronic gadgetry (['gædʒitri]n. 小配件, 小玩意).

② AT&T，美国电话电报公司。

By using a small radio transmitter① and a building full of special sensors, your desktop can be anywhere you are, not just at your workstation. At the press of a button, the computer closest to you in any room becomes your computer for as long as you need it. In addition to computers, the Cambridge researchers have designed the system to work for other devices, including phones and digital cameras.

2. Send out the Bat Signal

In order for a computer program to track its user, researchers had to develop a system that could locate both people and devices. The AT&T researchers came up with the ultrasonic location system. This location tracking system has three basic parts:

- Bats—small ultrasonic transmitters worn by users;
- Receivers—ultrasonic signal detectors embedded in ceiling;
- Central controller—coordinates the bats and receiver chains.

Users within the system will wear a bat, a small device that transmits a 48-bit code to the receivers in the ceiling. Bats also have an imbedded transmitter which allows it to communicate with the central controller using a bidirectional 433-MHz radio link.

Bats are 3 inches long (7.5 cm) by 1.4 inches wide (3.5 cm) by 0.6 inches thick (1.5 cm). These small devices are powered by a single 3.6-volt lithium thionyl chloride battery, which has a lifetime of six months. The devices also contain two buttons, two Light-Emitting Diodes (LEDs) and a piezoelectric speaker, allowing them to be used as ubiquitous input and output devices, and a voltage monitor to check the battery status(See Figure 4.2).

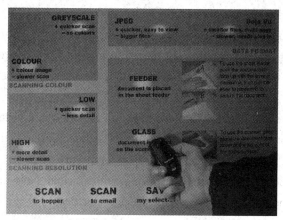

A smart poster will have buttons printed on to it that can be triggered by a bat.

Figure 4.2   Smart poster

---

① "Radio waves" transmit music, conversations, pictures and data <u>invisibly</u> ([inˈvizəbl]adv. 看不见地, 看不出地) through the air, often over millions of miles—it happens every day in thousands of different ways! Even though radio waves   are invisible and completely <u>undetectable</u> ([ˌʌndiˈtektəbl]adj. 无法觉察的, 无法发现的) to humans, they have totally changed society. Whether we are talking about a cell phone, a baby monitor, a <u>cordless phone</u> (无绳电话) or any one of the thousands of other wireless technologies, all of them use <u>radio waves</u> (无线电波) to communicate.

A bat will transmit an ultrasonic signal, which will be detected by receivers located in the ceiling approximately 4 feet (1.2 m) apart in a square grid. There are about 720 of these receivers in the 10,000-square-foot building (929 m^2) at the AT&T Labs in Cambridge. An object's location is found using trilateration, a position-finding technique that measures the objects distance in relation to three reference points.

If a bat needs to be located, the central controller sends the bat's ID over a radio link to the bat. The bat will detect its ID and send out an ultrasonic pulse. The central controller measures the time it took for that pulse to reach the receiver. Since the speed of sound through air is known, the position of the bat is calculated by measuring the speed at which the ultrasonic pulse reached three other sensors. This system provides a location accuracy of 1.18 inches (3 cm) throughout the Cambridge building.

By finding the position of two or more bats, the system can determine the orientation of a bat. The central controller can also determine which way a person is facing by analyzing the pattern of receivers that detected the ultrasonic signal and the strength of the signal.

3. In the Zone

With an ultrasonic location system in place, it's possible for any device fitted with a bat to become yours at the push of a button. Let's say the user leaves his workstation and enters another room. There's a phone in this room sitting on an unoccupied desk. That phone is now the user's phone and all of the user's phone calls are immediately redirected to that phone. If there is already someone using that phone, the central controller recognizes that and the person using the phone maintains possession of the phone.

The central controller creates a zone around every person and object within the location system. For example, if several cameras are placed in a room for videoconferences, the location system would activate the appropriate camera so that the user could be seen and move freely around the room.

When all the sensors and bats are in place, they are included in a virtual map of the building. The computer uses a spatial monitor to detect if a user's zone overlaps with the zone of a device. If the zones do overlap, then the user can become the temporary owner of the device.

If the ultrasonic location system is working with Virtual Network Computing (VNC)[①] software, there are some additional capabilities. Computer desktops can be created that actually follow their owners anywhere with in the system. Just by approaching any computer display in the building, the bat can enable the VNC desktop to appear on that display. This is handy if you want to leave your computer to show a coworker what you've been working on. Your desktop is simply teleported from your computer to your coworker's computer.

---

① In computing, Virtual Network Computing (VNC) is a graphical desktop sharing system that uses the RFB (Remote Frame Buffer，远程帧缓冲) protocol to remotely control another computer. It transmits the keyboard and mouse events from one computer to another, relaying the graphical screen updates back in the other direction, over a network.

4. Information Hoppers and Smart Posters

Once these zones are set up, computers on the network will have some interesting capabilities. The system will help us store and retrieve data in an "information hopper". This is a timeline of information that keeps track of when data is created. The hopper knows who created it, where they were and who they were with.

Think of the hopper as a ubiquitous filing clerk. It will change how we think of our computer filing systems. By using a digital camera that is connected to the network, a user's photographs are immediately stored in his or her timeline. Tape recorders could also send audio memos to the information hopper.

Two items of information created at the same time will be found at the same place on the timeline. The system knows who the user was with when he created the data, and the various timelines of the users working together. This way another timeline can be created to keep track of particular projects.

Another application that will come out of this ultrasonic location system is the smart poster. A conventional computer interface requires us to click on a button on our computer screen. In this new system, a button can be placed anywhere in your workplace, not just on the computer display. The idea behind smart posters is that a button can be a piece of paper that is printed out and stuck on a wall.

Smart posters will be used to control any device that is plugged into the network. The poster will know where to send a file and a user's preferences. Smart posters could also be used in advertising new services. To press a button on a smart poster, a user will simply place his or her bat on the smart poster button and click the bat. The system automatically knows who is pressing the poster's button. Posters can be created with several buttons on it.

Ultrasonic location systems will require us to think outside of the box. Traditionally, we have used our computer at work to store all of our files, and we may back up these files on a network server. This new ubiquitous network will enable all computers in a building to transfer ownership and store all of our files in a central timeline.

New Words

ubiquitous	[juːˈbikwitəs]	adj.	泛在的，到处存在的，普遍存在的
unchain	[ˈʌnˈtʃein]	vt.	解除，释放
transmitter	[trænzˈmitə]	n.	传导物，发报机，发射机
workstation	[ˈwəːksteiʃən]	n.	工作站
track	[træk]	n.	轨迹，跟踪，航迹，途径
		vt.	追踪
ultrasonic	[ˈʌltrəˈsɔnik]	adj.	超音速的，超声的
		n.	超声波

单词	音标	释义
bat	[bæt]	n. 蝙蝠，球棒
receiver	[riˈsiːvə]	n. 接收器，收信机
ceiling	[ˈsiːliŋ]	n. 天花板；最高限度
coordinate	[kəuˈɔːdinit]	n. 同等者，同等物，坐标(用复数)
		adj. 同等的，并列的
		vt. 调整，整理
bidirectional	[ˌbaidiˈrekʃənəl]	adj. 双向的
button	[ˈbʌtn]	n. 钮扣，按钮
piezoelectric	[paiˌiːzəuiˈlektrik]	adj. 压电的
status	[ˈsteitəs]	n. 情形，状况
trilateration	[traiˌlætəˈreiʃən]	n. 三边测量(术)
measure	[ˈmeʒə]	n. 量度器，量度标准，方法，测量，措施
		vt. 测量，测度，估量
pulse	[pʌls]	n. 脉冲
calculate	[ˈkælkjuleit]	vt. 计算，算出
accuracy	[ˈækjurəsi]	n. 精确性，正确度
unoccupied	[ʌnˈɔkjupaid]	adj. 没有人住的，无人占领的，空闲的
redirect	[ˈriːdiˈrekt]	vt. 使改道，使改变方向，重寄
zone	[zəun]	n. 地域，环带，圈
		vt. 环绕，使分成地带
		vi. 分成区
videoconference	[ˌvidiəuˈkɔnfərəns]	n. 视频会议
activate	[ˈæktiveit]	vt. 刺激，使活动
		vi. 有活力
appropriate	[əˈprəupriət]	adj. 适当的
overlap	[ˈəuvəˈlæp]	v. (与……)交叠
handy	[ˈhændi]	adj. 手边的，就近的，便利的，敏捷的，容易取得的
coworker	[kəˈwəːkə]	n. 合作者，同事
teleport	[ˈtelipɔːt]	vt. 传送
timeline	[ˈtaimlain]	n. 时间轴，时间线；大事年表
audio	[ˈɔːdiəu]	adj. 音频的，声频的，声音的
particular	[pəˈtikjulə]	n. 细节，详细
		adj. 特殊的，特别的，详细的，精确的
conventional	[kənˈvenʃnl]	adj. 惯例的，常规的，习俗的，传统的
preference	[ˈprefərəns]	n. 偏爱，优先选择
advertising	[ˈædvətaiziŋ]	n. 广告业，广告
		adj. 广告的
transfer	[trænsˈfəː]	n. & vt. 移动，传递，转移，转账，转让

ownership [ˈəunəʃip] n. 所有权，物主身份

Phrases

put a new spin on	对……做出新的解释
ubiquitous networking system	泛在网系统
radio transmitter	无线电广播发射机
as long as	只要，在……的时候
digital camera	数码相机
come up with	提出，想出，赶上
ultrasonic location system	超声波定位系统
ultrasonic transmitter	超声发射机，超声波发射器
ultrasonic signal detector	超声检测信号探测器
lithium thionyl chloride battery	锂亚硫酰氯原电池
square grid	方格网，方栅
in relation to …	关于，涉及，与……相比
set up	设立，竖立，架起，升起
information hopper	信息料斗，信息仓，信息存储池
keep track of	记录，持续追踪；与……保持联系
filing clerk	文件管理员
tape recorder	磁带录音机
smart poster	智能海报
print out	打印出，印出，显示
plug into	把(电器)插头插入，接通
think outside of the box	跳出时间框框，创意思维，打破常规
back up	备份

Abbreviations

PDA (Personal Digital Assistant) 个人数字助理
LED (Light-Emitting Diode) 发光二极管
VNC (Virtual Network Computing) 虚拟网络计算

Exercises

I. Answer the following questions according to the text.

1. What are the researchers doing in addition to taking the hardware with you?
2. How can your desktop be anywhere you are, not just at your workstation?
3. How many basic parts does the ultrasonic location system have? What are they?

4. What is the size of the bats?

5. How is the position of the bat calculated since the speed of sound through air is known?

6. What can the central controller determine?

7. What does the computer use a spatial monitor to do? What happens if they do?

8. What will the system help us?

9. What can a user do by using a digital camera that is connected to the network?

10. What will smart posters be used to do?

II. Translate the following terms or phrases from English into Chinese and vice versa.

1. piezoelectric   1. _____
2. workstation   2. _____
3. redirect   3. _____
4. 方格网，方栅   4. _____
5. pulse   5. _____
6. ubiquitous   6. _____
7. n. 接收器，收信机   7. _____
8. back up   8. _____
9. bidirectional   9. _____
10. information hopper   10. _____

III. Fill in the blanks with the words given below.

hardware   accommodate   operating   LAN   configure
adapters   multiplayer   package   subnets   sharing

◆◆ LAN ◆◆

1. Definition

A Local Area Network (LAN) supplies networking capability to a group of computers in close proximity to each other such as in an office building, a school, or a home. A LAN is useful for __1__ resources like files, printers, games or other applications. A __2__ in turn often connects to other LANs, and to the Internet or other WAN.

Most Local Area Networks are built with relatively inexpensive __3__ such as Ethernet cables, network __4__, and hubs. Wireless LAN and other more advanced LAN hardware options also exist.

Specialized __5__ system software may be used to __6__ a Local Area Network. For example, most flavors of Microsoft Windows provide a software __7__ called Internet Connection Sharing (ICS) that supports controlled access to LAN resources.

The term LAN party refers to a __8__ gaming event where participants bring their own computers and build a temporary LAN.

2. Examples

The most common type of local area network is an Ethernet LAN. The smallest home LAN can have exactly two computers; a large LAN can ___9___ many thousands of computers. Many LANs are divided into logical groups called ___10___. An Internet Protocol (IP) "Class A" LAN can in theory accommodate more than 16 million devices organized into subnets.

IV. Translate the following passage from English to Chinese.

✧✧ WAN ✧✧

A Wide Area Network (WAN) is a telecommunication network that covers a broad area (i.e., any network that links across metropolitan, regional, or national boundaries). Business and government entities utilize WANs to relay data among employees, clients, buyers, and suppliers from various geographical locations. In essence this mode of telecommunication allows a business to effectively carry out its daily function regardless of location.

This is in contrast with Personal Area Networks (PANs), Local Area Networks (LANs), Campus Area Networks (CANs), or Metropolitan Area Networks (MANs) which are usually limited to a room, building, campus or specific metropolitan area respectively.

VPNs

As a business grows, it might expand to multiple shops or offices across the country and around the world. To keep things running efficiently, the people working in those locations need a fast, secure and reliable way to share information across computer networks. In addition, traveling employees like salespeople need an equally secure and reliable way to connect to their business's computer network from remote locations.

One popular technology to accomplish these goals is a VPN (Virtual Private Network). A VPN is a private network that uses a public network (usually the Internet) to connect remote sites or users together. The VPN uses "virtual" connections routed through the Internet from the business's private network to the remote site or employee. By using a VPN, businesses ensure security—anyone intercepting the encrypted data can't read it.

VPN was not the first technology to make remote connections. Several years ago, the most common way to connect computers between multiple offices was by using a leased line. Leased lines, such as ISDN (Integrated Services Digital Network[①], 128 Kbps), are private network

---

① Integrated Services Digital Network (ISDN) is a set of communications standards for simultaneous digital transmission of voice, video, data, and other network services over the traditional circuits of the public switched telephone network (公共交换电话网).

connections that a telecommunications company could lease to its customers. Leased lines provided a company with a way to expand its private network beyond its immediate geographic area. These connections form a single Wide Area Network (WAN) for the business. Though leased lines are reliable and secure, the leases are expensive, with costs rising as the distance between offices increases.

Today, the Internet is more accessible than ever before, and Internet Service Providers (ISPs) continue to develop faster and more reliable services at lower costs than leased lines. To take advantage of this, most businesses have replaced leased lines with new technologies that use Internet connections without sacrificing performance and security. Businesses started by establishing intranets, which are private internal networks designed for use only by company employees. Intranets enabled distant colleagues to work together through technologies such as desktop sharing. By adding a VPN, a business can extend all its intranet's resources to employees working from remote offices or their homes.

1. Analogy: Each LAN is an Island

Imagine that you live on an island in a huge ocean. There are thousands of other islands all around you, some very close and others farther away. The common means of travel between islands is via ferry. Traveling on the ferry means that you have almost no privacy: Other people can see everything you do.

Let's say that each island represents a private Local Area Network (LAN) and the ocean is the Internet. Traveling by ferry is like connecting to a Web server or other device through the Internet. You have no control over the wires and routers that make up the Internet, just like you have no control over the other people on the ferry. This leaves you susceptible to security issues if you're trying to connect two private networks using a public resource.

Continuing with our analogy, your island decides to build a bridge to another island so that people have an easier, more secure and direct way to travel between the two islands. It is expensive to build and maintain the bridge, even if the islands are close together. However, the need for a reliable, secure path is so great that you do it anyway. Your island would like to connect to yet another island that is much farther away, but decides that the costs are simply too much to bear.

This scenario represents having a leased line. The bridges (leased lines) are separate from the ocean (Internet), yet are able to connect the islands (LANs[①]). Companies who choose this option do so because of the need for security and reliability in connecting their remote offices. However, if the offices are very far apart, the cost can be prohibitively high—just like trying to build a bridge that spans a great distance.

---

① Local Area Network (LAN) is a computer network that interconnects computers in a limited area such as a home, school, computer laboratory, or office building. The defining characteristics of LANs, in contrast to Wide Area Networks (WANs), include their usually higher <u>data-transfer rates</u> (数据传输率), smaller geographic area, and lack of a need for leased telecommunication lines.

So how does a VPN fit in? Using our analogy, suppose each inhabitant on your island has a small submarine. Let's assume that each submarine has these amazing properties:
• It's fast.
• It's easy to take with you wherever you go.
• It's able to completely hide you from any other boats or submarines.
• It's dependable.
• It costs little to add additional submarines to your fleet once you've purchased the first one.

Although they're traveling in the ocean along with other traffic, the people could travel between islands whenever they wanted to with privacy and security. That's essentially how a VPN works. Each remote member of your network can communicate in a secure and reliable manner using the Internet as the medium to connect to the private LAN. A VPN can grow to accommodate more users and different locations much more easily than a leased line. In fact, scalability is a major advantage that VPNs have over leased lines. Moreover, the distance doesn't matter, because VPNs can easily connect multiple geographic locations worldwide.

2. What Makes a VPN?

A VPN's purpose is providing a secure and reliable private connection between computer networks over an existing public network, typically the Internet. Before looking at the technology that makes a VPN possible, let's consider all the benefits and features a business should expect in a VPN.

A well-designed VPN provides a business with the following benefits:
• Extended connections across multiple geographic locations without using a leased line;
• Improved security for exchanging data;
• Flexibility for remote offices and employees to use the business intranet over an existing Internet connection as if they're directly connected to the network;
• Savings in time and expense for employees to commute if they work from virtual workplaces[①];
• Improved productivity for remote employees.

A business might not require all these benefits from its VPN, but it should demand the following essential VPN features:
• Security—The VPN should protect data while it's traveling on the public network. If intruders attempt to capture the data, they should be unable to read or use it.
• Reliability—Employees and remote offices should be able to connect to the VPN with no trouble at any time (unless hours are restricted), and the VPN should provide the same quality of

---

① A virtual workplace is a workplace that is not located in any one physical space. Rather, several workplaces are technologically connected (via the Internet) without regard to geographic boundaries. Employees are thus able to interact and work with one another in a collaborated environment regardless of where they are in the world. A virtual workplace decreases <u>unnecessary costs</u> (多余的成本，不必要的成本) by integrating technology processes, people processes, and online processes.

connection for each user even when it is handling its maximum number of simultaneous connections.

• Scalability—As a business grows, it should be able to extend its VPN services to handle that growth without replacing the VPN technology altogether.

One interesting thing to note about VPNs is that there are no standards about how to set them up. This article covers network, authentication and security protocols that provide the features and benefits listed above. It also describes how a VPN's components work together. If you're establishing your own VPN, though, it's up to you to decide which protocols and components to use and to understand how they work together.

3. Two common type of VPN

3.1 Remote-access VPN

A remote-access VPN allows individual users to establish secure connections with a remote computer network. Those users can access the secure resources on that network as if they were directly plugged in to the network's servers. An example of a company that needs a remote-access VPN is a large firm with hundreds of salespeople in the field. Another name for this type of VPN is Virtual Private Dial-up Network (VPDN), acknowledging that in its earliest form, a remote-access VPN required dialing in to a server using an analog telephone system.

There are two components required in a remote-access VPN. The first is a Network Access Server[①] (NAS, usually pronounced "nazz" conversationally), also called a media gateway or a Remote-Access Server (RAS)[②]. A NAS might be a dedicated server, or it might be one of multiple software applications running on a shared server. It's a NAS that a user connects to from the Internet in order to use a VPN. The NAS requires that user to provide valid credentials to sign in to the VPN. To authenticate the user's credentials, the NAS uses either its own authentication process or a separate authentication server running on the network.

The other required component of remote-access VPNs(See Figure 4.3) is client software. In other words, employees who want to use the VPN from their computers require software on those computers that can establish and maintain a connection to the VPN. Most operating systems[③] today have built-in software that can connect to remote-access VPNs, though some VPNs might

---

① A Network Access Server (NAS) is a single point of access to a remote resource.
② Remote Access Services (RAS) refers to any combination of hardware and software to enable the remote access tools or information that typically reside on a network of IT devices. A RAS server is a specialized computer which aggregates (['ægrigeit]v. 聚集, 集合) multiple communication channels together. Because these channels are bidirectional ([,baidi'rekʃənəl]adj. 双向的), two models emerge: Multiple entities connecting to a single resource, and a single entity connecting to multiple resources. Both of these models are widely used. Both physical and virtual resources can be provided through a RAS server: centralized computing can provide multiple users access to a remote virtual operating system.
③ An Operating System (OS) is a set of programs that manage computer hardware resources and provide common services for application software. The operating system is the most important type of system software in a computer system.

require users to install a specific application instead. The client software sets up the tunneled connection to a NAS, which the user indicates by its Internet address. The software also manages the encryption required to keep the connection secure.

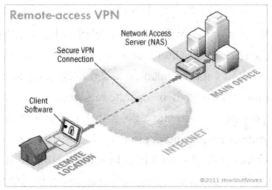

A remote-access VPN connection allows an individual user to connect to a private business network from a remote location using a laptop or desktop computer connected to the Internet.

Figure 4.3　Remote-access VPN

Large corporations or businesses with knowledgeable IT staff typically purchase, deploy and maintain their own remote-access VPNs. Businesses can also choose to outsource their remote-access VPN services through an Enterprise Service Provider (ESP). The ESP sets up a NAS for the business and keeps that NAS running smoothly.

A remote-access VPN is great for individual employees, but what about entire branch offices with dozens or even hundreds of employees? Next, we'll look at another type of VPN used to keep businesses connected LAN-to-LAN.

3.2　Site-to-site VPN

A site-to-site VPN allows offices in multiple fixed locations to establish secure connections with each other over a public network such as the Internet. Site-to-site VPN extends the company's network, making computer resources from one location available to employees at other locations. An example of a company that needs a site-to-site VPN is a growing corporation with dozens of branch offices around the world.(See Figure 4.4)

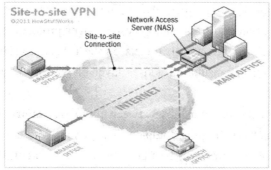

A site-to-site VPN connection lets branch offices use the Internet as a conduit for accessing the main office's intranet.

Figure 4.4　Site-to-site VPN

There are two types of site-to-site VPNs:

• Intranet-based—If a company has one or more remote locations that they wish to join in a single private network, they can create an intranet VPN to connect each separate LAN to a single WAN.

• Extranet-based—When a company has a close relationship with another company (such as a partner, supplier or customer), it can build an extranet VPN that connects those companies' LANs. This extranet VPN allows the companies to work together in a secure, shared network environment while preventing access to their separate intranets.

Even though the purpose of a site-to-site VPN is different from that of a remote-access VPN, it could use some of the same software and equipment. Ideally, though, a site-to-site VPN should eliminate the need for each computer to run VPN client software as if it were on a remote-access VPN. Dedicated VPN client equipment can accomplish this goal in a site-to-site VPN.

Now that you know the two types of VPNs, let's look at how your data is kept secure as it travels across a VPN.

4. Keeping VPN Traffic in the Tunnel

Most VPNs rely on tunneling to create a private network that reaches across the Internet. Tunneling is the process of placing an entire packet of data file within another packet before it's transported over the Internet. That outer packet protects the contents from public view and ensures that the packet moves within a virtual tunnel.

This layering of packets is called encapsulation. Computers or other network devices at both ends of the tunnel, called tunnel interfaces, can encapsulate outgoing packets and reopen incoming packets. Users (at one end of the tunnel) and IT personnel (at one or both ends of the tunnel) configure the tunnel interfaces they're responsible for to use a tunneling protocol. Also called an encapsulation protocol, a tunneling protocol is a standardized way to encapsulate packets.

The purpose of the tunneling protocol is to add a layer of security that protects each packet on its journey over the Internet. The packet is traveling with the same transport protocol it would have used without the tunnel; this protocol defines how each computer sends and receives data over its ISP. Each inner packet still maintains the passenger protocol, such as Internet Protocol (IP)① or AppleTalk, which defines how it travels on the LANs at each end of the tunnel. The tunneling protocol used for encapsulation adds a layer of security to protect the packet on its journey over the Internet.

To better understand the relationships between protocols, think of tunneling as having a computer delivered to you by a shipping company. The vendor who is sending you the computer packs the computer (passenger protocol) in a box (tunneling protocol). Shippers then place that

---

① The Internet Protocol (IP) is the principal communications protocol used for relaying datagrams (packets) across an internetwork using the Internet Protocol Suite. <u>Responsible for</u> (为……负责) routing packets across network boundaries, it is the primary protocol that establishes the Internet.

box on a shipping truck (transport protocol) at the vendor's warehouse (one tunnel interface). The truck (transport protocol) travels over the highways (Internet) to your home (the other tunnel interface) and delivers the computer. You open the box (tunneling protocol) and remove the computer (passenger protocol).

New Words

expand	[iks'pænd]	v. 扩展
efficiently	[i'fiʃəntli]	adv. 有效率地，有效地
reliable	[ri'laiəbl]	adj. 可靠的，可信赖的
share	[ʃɛə]	n. 共享，参与，一份，部分，份额，参股
		vt. 分享，均分，共有，分配
		vi. 分享
salespeople	['seilz,pi:pl]	n. 售货员，店员
equally	['i:kwəli]	adv. 相等地，平等地，公平地
accomplish	[ə'kɔmpliʃ]	vt. 完成，达到，实现
site	[sait]	n. 站点
intercept	[,intə'sept]	vt. 中途阻止，截取
encrypt	[in'kript]	v. 加密，将……译成密码
accessible	[ək'sesəbl]	adj. 可访问的，易接近的，可到达的
sacrifice	['sækrifais]	n. & v. 牺牲
performance	[pə'fɔ:məns]	n. 履行，执行，性能
internal	[in'tə:nl]	adj. 内在的，内部的
colleague	['kɔli:g]	n. 同事，同僚
resource	[ri'sɔ:s]	n. 资源
analogy	[ə'nælədʒi]	n. 类似，类推
ferry	['feri]	n. 摆渡，渡船，渡口
		vt. 渡运，(乘渡船)渡过，运送
		vi. 摆渡
maintain	[men'tein]	vt. 维持，维修
bear	[bɛə]	v. 负担，忍受，带给
span	[spæn]	n. 跨度，跨距，范围
		v. 横越
submarine	['sʌbməri:n]	n. 潜水艇，潜艇
dependable	[di'pendəbl]	adj. 可靠的
fleet	[fli:t]	n. 舰队，港湾
		adj. 快速的，敏捷的
purchase	['pə:tʃəs]	vt. & n. 买，购买
essentially	[i'senʃəli]	adv. 本质上，本来

worldwide	[ˈwəːldwaid]	adj. 全世界的
commute	[kəˈmjuːt]	v. 交换，抵偿
productivity	[ˌprɔdʌkˈtiviti]	n. 生产力
handling	[ˈhændliŋ]	n. 处理
		adj. 操作的
attempt	[əˈtempt]	n. 努力，尝试，企图
		vt. 尝试，企图
standard	[ˈstændəd]	n. 标准，规格
		adj. 标准的，权威，第一流的
dedicated	[ˈdedikeitid]	adj. 专注的
credential	[kriˈdenʃəl]	n. 信任状
tunnel	[ˈtʌnl]	n. 通道，隧道，地道
knowledgeable	[ˈnɔlidʒəbl]	adj. 知识渊博的，有见识的
staff	[stɑːf]	n. 全体职员
outsource	[ˈautsɔːs]	vt. 把……外包
		vi. 外包
smoothly	[ˈsmuːðli]	adv. 平稳地
partner	[ˈpɑːtnə]	n. 合伙人，股东，伙伴
		vt. 与……合伙，组成一对
		vi. 做伙伴，当助手
supplier	[səˈplaiə]	n. 供应者，供应商，供给者
customer	[ˈkʌstəmə]	n. 消费者
prevent	[priˈvent]	v. 防止，预防
encapsulation	[inˌkæpsjuˈleiʃən]	n. 封装，包装
encapsulate	[inˈkæpsjuleit]	v. 封装
outgoing	[ˈautgəuiŋ]	n. 外出，开支，流出
reopen	[ˈriːˈəupən]	v. 重开，再开始
shipper	[ˈʃipə]	n. 托运人，发货人
warehouse	[ˈwɛəhaus]	n. 仓库，货栈，大商店
		vt. 存入仓库
truck	[trʌk]	n. 卡车，交易，交换
		vt. 交易，交往，以卡车运输
		vi. 驾驶卡车，以物易物
highway	[ˈhaiwei]	n. 公路，大路，高速公路

Phrases

| leased line | 租用线，专用线 |
| telecommunications company | 电信公司，通信公司 |

at lower costs	以较低成本，以较低价格
take advantage of	利用
replaced … with …	用……替换……
have no control over	不能控制
make up	连成，弥补，缝制，整理
public resource	公共资源
in the field	在野外
analog telephone system	模拟电话系统
sign in	签到，签收
client software	客户软件
dozens of	许多的
branch office	分支机构
have a close relationship with	有密切的联系
tunnel interface	隧道接口
be responsible for	为……负责，形成……的原因
transport protocol	传输协议
passenger protocol	乘客协议
deliver to	转交，交付，传达
shipping company	轮船公司
shipping truck	运输卡车

Abbreviations

VPN (Virtual Private Network)	虚拟专用网络，虚拟个人网络
ISDN (Integrated Services Digital Network)	综合业务服务网，综合服务数字网
VPDN (Virtual Private Dial-up Network)	虚拟专用拨号网
NAS (Network Access Server)	网络访问服务器
RAS (Remote-Access Server)	远程访问服务器
IT (Information Technology)	信息技术
ESP (Enterprise Service Provider)	企业服务提供商

Exercises

I. Fill in the blanks with the information given in the text.

1. What does VAN stand for? What is it?

2. What was the most common way to connect computers between multiple offices several years ago?

3. What is a major advantage that VPNs have over leased lines? Why doesn't distance matter for VANs?

4. What is a VAN's purpose?
5. What does a remote-access VPN allow individual users to do?
6. How many components does a remote-access VPN require?
7. What does a site-to-site VPN allow offices in multiple fixed locations to do?
8. How many types of site-to-site VPNs are there? What are they?
9. What is tunneling?
10. What is the purpose of the tunneling protocol?

II. Translate the following terms or phrases from English into Chinese and vice versa.

1. expand
2. v. 封装
3. v. 加密，将……译成密码
4. tunnel
5. performance
6. tunnel interface
7. maintain
8. transport protocol
9. leased line
10. outsource

Text A 参考译文

泛在网是如何工作的？

1. 泛在网的工作方式简介

移动计算设备已经改变了我们对计算的看法。便携计算机和个人数字助理(如图 4.1 所示，图略)已经把我们从桌面计算机中解放出来。AT&T 剑桥实验室的一群研究人员正准备提出移动计算的新思路。除了为你提供硬件之外，他们正在设计一种泛在网系统，让应用程序随时随地跟随你。

通过使用一个小的无线电发送器和特定的传感器，你的桌面可以出现在你所在的任何位置，而不只是在你的工作站上。如果你需要，则当按下任何一个按钮时，你附近任何一个房间的计算机都可以成为你的计算机。除了计算机之外，剑桥的研究者也已经将该系统设计成可以用于其他设备(包括电话和数码相机)。

2. 发送短棒设备信号

为了让计算机程序能跟踪它的用户，研究者必须开发一个可以定位人和设备的系统。AT&T 的研究人员提出了超声波定位系统。这个定位跟踪系统有三个基本组成部分：

- 短棒设备——用户携带的小超声波发射器；
- 接收器——嵌入在天花板上的超声波信号探测器；
- 中心控制器——配合短棒和接收链。

该系统中的用户携带一个短棒设备，即一个把 48 位编码传输给天花板上接收器的小型设备。短棒设备也具有一个嵌入式发射器，使其自身可以通过使用 433 MHz 双向无线链路来与中心处理器通信。

短棒设备有 3 英寸长(7.5 cm)、1.4 英寸宽(3.5 cm)、0.6 英寸厚(1.5 cm)。这些小设备使用单一的 3.6 V 的氯化亚硫酰锂电池，电池寿命为 6 个月。该设备还包含两个按钮、两个发光二极管以及一个压电扬声器(可以用作泛在输入和输出设备)，还有一个可以检测电池状态的电压监控器(如图 4.2 所示，图略)。

短棒设备会传输超声波信号，该信号可以被天花板上的接收器探测到，这些接收器布置在大约 4 英尺(1.2 m)的范围内。剑桥 AT&T 实验室 10000 平方英尺($929\ m^2$)的范围内布置了大约 720 个接收器。可以通过三边定位法来确定物体的位置，该方法使用物体与三个参考点的距离来定位。

如果需要对短棒设备定位，则中心处理器通过无线链路给该短棒设备发送其 ID。该短棒设备探测到其 ID 并发出超声波脉冲，然后中心控制器测量该脉冲到达接收器所用的时间。因为声音在空气中的传播速度是已知的，所以可以通过计算超声波脉冲到达其他三个传感器的时间来得到该短棒设备的位置。在剑桥的大楼中，该系统提供了 1.18 英寸(3 cm) 的定位精度。

通过寻找两个或多个短棒设备的位置，该系统可以确定一个短棒设备的方向。通过分析接收器检测到的超声波信号及信号的强度，中心处理器也可以确定一个人的走向。

3. 在区域中

有了超声波定位系统，只要按一下按钮，装有短棒设备的任何装置都可能为你所用。比如说，用户离开其工作站进入另一房间，该房间的空桌子上有一个电话，那该电话现在就是这个用户的电话了，因为所有呼叫该用户的电话都会立即转接到这个电话上。如果已经有人使用了这个电话，那么中心处理器识就会别出来，且已使用该电话的人将继续拥有该电话。

中心处理器为定位系统中的每个人和物体建立了一个区域。例如，如果为视频会议在一个房间安置了几个照相机，定位系统将激活适当的照相机以便可以看到房间中的用户并允许其随便移动。

若所有的传感器和短棒设备都放置适当，则它们可以被包含在大楼的虚拟地图中。计算机使用空间监控器来探测用户的区域是否与设备的区域重叠。如果区域确实重叠，则用户可以临时拥有该设备。

如果超声波定位系统和虚拟网络计算机软件一起工作，则会有某些附加性能，如该系统中任何地方的用户都可以建立自己的计算机桌面。只要靠近该大楼中的任一计算机显示器，该短棒设备就可以在屏幕上显示 VNC 桌面。如果你想离开你的计算机去告诉你的工友你在做什么，这很方便，只需将你的桌面从你的计算机移到你工友的计算机上就行了。

4. 信息存储池和智能海报

一旦建立了这些区域，网络中的计算机会有一些有趣的性能。该系统将帮助我们在"信息存储池"中保存和检索数据。信息存储池是一个信息时间轴，可以跟踪已经建立的数据。该存储池知道谁建立了它、他们在哪里以及和谁在一起。

可以把该存储池看做泛在的文件管理员，它将改变我们对计算机文件系统的看法。通过使用连接到该网络上的数码相机，用户的照片立即存储在他或她的时间轴上。录音机也可以发送音频备忘录给信息存储池。

同时建立的两个信息项将出现在时间轴上的相同位置。当用户建立数据时，系统知道用户是谁，以及用户用到的多个时间轴。这样可以建立另外的时间轴来跟踪特定的项目。

这种超声波定位系统能提供的另一应用是智能海报。常规的计算机界面需要我们点击屏幕上的按钮。但在这种新系统中，按钮可以放置在工作场所的任何地方，不只是在计算机屏幕上。智能海报的设计思想是可将该按钮印在纸上并粘贴在墙上。

智能海报可以用于控制联入网络的任何设备。海报会知道发送文件的地方和用户参数。智能海报也可以用来发布最新服务的广告。要按智能海报上的按钮，用户只需把他/她的短棒设备对准智能海报的按钮点击即可。系统自动地知道谁按下了海报的按钮。海报上可以放置数个按钮。

超声波定位系统要求我们跳出思维的框框。传统意义上，我们使用自己的计算机存储我们所有的文件，并把这些文件备份到网络服务器上。而新的泛在网能够使一个大楼内的所有计算机可相互移交文件的拥有权，并把全部文件存储在一个中心时间轴上。

Unit 5 Barcode

Barcode

A barcode is an optical machine-readable representation of data, which shows data about the object to which it attaches. Originally barcodes represented data by varying the widths and spacings of parallel lines, and may be referred to as linear or 1 Dimensional (1D). Later they evolved into rectangles, dots, hexagons and other geometric patterns in 2 Dimensions (2D). Although 2D systems use a variety of symbols, they are generally referred to as barcodes as well. Barcodes originally were scanned by special optical scanners called barcode readers; later, scanners and interpretive software became available on devices including desktop printers and smartphones.

The first use of barcodes was to label railroad cars, but they were not commercially successful until they were used to automate supermarket checkout systems, a task for which they have become almost universal(See Figure 5.1). Their use has spread to many other tasks that are generically referred to as Automatic Identification and Data Capture (AIDC)①. The very first scanning of the now ubiquitous Universal Product Code (UPC)② barcode was on a pack of Wrigley Company chewing gum in June 1974.

Figure 5.1 A UPC-A barcode symbol

---

① Automatic Identification and Data Capture (AIDC) refers to the methods of automatically identifying objects, collecting data about them, and entering that data directly into computer systems (i.e. without human involvement). Technologies typically considered as part of AIDC include bar codes, Radio Frequency IDentification (RFID), biometrics ([,baiəu'metriks]n.生物识别技术), magnetic stripes (磁条), Optical Character Recognition (OCR，光学字符识别), smart cards, and voice recognition (语音识别). AIDC is also commonly referred to as "Automatic Identification", "Auto-ID", and "Automatic Data Capture".

② The Universal Product Code (UPC) is a barcode symbology (i.e., a specific type of barcode). Its most common form, the UPC-A, consists of 12 numerical digits, which are uniquely assigned to each trade item.

Other systems have made inroads in the AIDC market, but the simplicity, universality and low cost of barcodes has limited the role of these other systems until the first decade of the 21st century, over 40 years after the introduction of the commercial barcode, came the introduction of technologies such as Radio Frequency IDentification (RFID).

1. Use

Barcodes such as the UPC have become a ubiquitous element of modern civilization, as evidenced by their enthusiastic adoption by stores around the world; almost every item other than fresh produce from a grocery store, department store, and mass merchandiser has a UPC barcode on it. This helps track items and also reduces instances of shoplifting involving price tag swapping, although shoplifters can now print their own barcodes. In addition, retail chain membership cards (issued mostly by grocery stores and specialty "big box" retail stores such as sporting equipment, office supply, or pet stores) use bar codes to uniquely identify consumers, allowing for customized marketing and greater understanding of individual consumer shopping patterns. At the point of sale, shoppers can get product discounts or special marketing offers through the address or E-mail address provided at registration.

Barcodes can allow for the organization of large amounts of data. They are widely used in the healthcare and hospital settings, ranging from patient identification (to access patient data, including medical history, drug allergies, etc.) to medication management. They are also used to facilitate the separation and indexing of documents that have been imaged in batch scanning applications, track the organization of species in biology, and integrate with in-motion check weighers to identify the item being weighed in a conveyor line for data collection.

They can also be used to keep track of objects and people; they are used to keep track of rental cars, airline luggage, nuclear waste, registered mail, express mail and parcels. Barcoded tickets allow the holder to enter sports arenas, cinemas, theatres, fairgrounds, and transportation, and are used to record the departure and arrival of vehicles from rental facilities etc. This can allow proprietors to identify duplicate or fraudulent tickets more easily. Barcodes are widely used in shop floor control applications software where employees can scan work orders and track the time spent on a job.

Some 2D barcodes embed a hyperlink to a web page[①]. A capable cellphone might be used to read the pattern and browse the linked website, which can help a shopper find the best price for an item in the vicinity. Since 2005, airlines use an IATA-standard 2D barcode on boarding passes, and since 2008 2D barcodes sent to mobile phones enable electronic boarding passes.

---

① A web page or webpage is a document or information resource that is suitable for the World Wide Web and can be accessed through a web browser and displayed on a monitor or mobile device. This information is usually in HTML (Hypertext Markup Language, 超文本标识语言) or XHTML (eXtensible HyperText Markup Language, 可扩展超文本标识语言) format, and may provide navigation to other web pages via hypertext links (超链接). Web pages frequently subsume ([sʌbˈsjuːm]v. 包含) other resources such as style sheets (样式表), scripts ([skrɪpt]n. 脚本) and images into their final presentation.

2. Symbologies

The mapping between messages and barcodes is called a symbology. The specification of a symbology includes the encoding of the single digits/characters of the message as well as the start and stop markers into bars and space, the size of the quiet zone required to be before and after the barcode as well as the computation of a checksum.

Linear symbologies can be classified mainly by two properties:

• Continuous vs. discrete: characters in continuous symbologies usually abut, with one character ending with a space and the next beginning with a bar, or vice versa. Characters in discrete symbologies begin and end with bars; the intercharacter space is ignored, as long as it is not wide enough to look like the code ends.

• Two-width vs. many-width: bars and spaces in two-width symbologies are wide or narrow; the exact width of a wide bar has no significance as long as the symbology requirements for wide bars are adhered to (usually two to three times wider than a narrow bar). Bars and spaces in many-width symbologies are all multiples of a basic width called the module; most such codes use four widths of 1, 2, 3 and 4 modules.

Some symbologies use interleaving. The first character is encoded using black bars of varying width. The second character is then encoded by varying the width of the white spaces between these bars. Thus characters are encoded in pairs over the same section of the barcode.

The most common among the many 2D symbologies are matrix codes, which feature square or dot-shaped modules arranged on a grid pattern. 2D symbologies also come in circular and other patterns and may employ steganography, hiding modules within an image.

Linear symbologies are optimized for laser scanners, which sweep a light beam across the barcode in a straight line, reading a slice of the barcode light-dark patterns. Stacked symbologies are also optimized for laser scanning, with the laser making multiple passes across the barcode.

In the 1990s development of Charge Coupled Device (CCD)① imagers to read barcodes was pioneered by Welch Allyn. Imaging does not require moving parts, as a laser scanner does. In 2007, linear imaging had begun to supplant laser scanning as the preferred scan engine for its performance and durability.

2D symbologies cannot be read by a laser as there is typically no sweep pattern that can encompass the entire symbol. They must be scanned by an image-based scanner employing a CCD or other digital camera sensor technology.

3. Scanners (barcode readers)

The earliest, and still the cheapest, barcode scanners are built from a fixed light and a single

---

① A Charge Coupled Device (CCD) is a device for the movement of electrical charge, usually from within the device to an area where the charge can be manipulated, for example conversion into a digital value. This is achieved by "shifting" the signals between stages within the device one at a time. CCDs move charge between capacitive bins ([bin]n. 箱柜) in the device, with the shift allowing for the transfer of charge between bins.

photosensor that is manually "scrubbed" across the barcode.

Barcode scanners can be classified into three categories based on their connection to the computer. The older type is the RS-232 barcode scanner. This type requires special programming for transferring the input data to the application program. "Keyboard interface scanners" connect to a computer using a PS/2 or AT keyboard-compatible adaptor cable. The barcode's data is sent to the computer as if it had been typed on the keyboard. Like the keyboard interface scanner, USB scanners are easy to install and do not need custom code for transferring input data to the application program.

4. Benefits

In point-of-sale management, barcode systems can provide detailed up-to-date information on the business, accelerating decisions and with more confidence. For example:

- Fast-selling items can be identified quickly and automatically reordered.
- Slow-selling items can be identified, preventing inventory build-up.
- The effects of merchandising changes can be monitored, allowing fast-moving, more profitable items to occupy the best space.
- Historical data can be used to predict seasonal fluctuations very accurately.
- Items may be repriced on the shelf to reflect both sale prices and price increases.
- This technology also enables the profiling of individual consumers, typically through a voluntary registration of discount cards. While pitched as a benefit to the consumer, this practice is considered to be potentially dangerous by privacy advocates.

Besides sales and inventory tracking, barcodes are very useful in logistics.

- When a manufacturer packs a box for shipment, a Unique Identifying Number (UID) can be assigned to the box.
- A database can link the UID to relevant information about the box; such as order number, items packed, quantity packed, destination, etc.
- The information can be transmitted through a communication system such as Electronic Data Interchange (EDI)[①] so the retailer has the information about a shipment before it arrives.
- Shipments that are sent to a Distribution Center (DC)[②] are tracked before forwarding. When the shipment reaches its final destination, the UID gets scanned, so the store knows the shipment's source, contents, and cost.

---

① Electronic Data Interchange (EDI) is the structured transmission of data between organizations by electronic means (电子化方法). It is used to transfer electronic documents or business data from one computer system to another computer system, i.e. from one trading partner to another trading partner without human intervention. It is more than mere E-mail. For instance, organizations might replace bills of lading and even cheques with appropriate EDI messages. It also refers specifically to a family of (一系列) standards.

② A distribution center for a set of products is a warehouse or other specialized building, often with refrigeration ([riˌfridʒəˈreiʃən]n. 冷藏) or air conditioning, which is stocked with products (goods) to be redistributed to retailers, to wholesalers, or directly to consumers. A distribution center is a principal part, the order processing element, of the entire order fulfillment process.

Barcode scanners are relatively low cost and extremely accurate compared to key-entry, with only about 1 substitution error in 15,000 to 36 trillion characters entered. The exact error rate depends on the type of barcode.

New Words

optical	[ˈɔptikəl]	adj. 光学的
attach	[əˈtætʃ]	v. 配属，隶属于
originally	[əˈridʒənəli]	adv. 最初，原先
represent	[ˌriːpriˈzent]	vt. 表现，描绘
spacing	[ˈspeisiŋ]	n. 间隔，间距
parallel	[ˈpærəlel]	adj. 平行的，相同的，类似的，并联的
		n. 平行线，类似，相似物
		v. 相应，平行
linear	[ˈliniə]	adj. 线的，直线的，线性的
rectangle	[ˈrektæŋgl]	n. 长方形，矩形
hexagon	[ˈheksəgən]	n. 六角形，六边形
scan	[skæn]	v. 浏览，扫描
		n. 扫描
scanner	[ˈskænə]	n. 扫描器
interpretive	[inˈtəːpritiv]	adj. 作为说明的，解释的
label	[ˈleibl]	n. 标签，签条，商标，标志
		vt. 贴标签于，指……为，分类，标注
automate	[ˈɔːtəmeit]	v. 使自动化，自动操作
checkout	[ˈtʃekaut]	n. 结账台；检验，校验
universal	[ˌjuːniˈvəːsəl]	adj. 普遍的，全体的，通用的，宇宙的，世界的
simplicity	[simˈplisiti]	n. 简单，简易
universality	[ˌjuːnivəˈsæliti]	n. 普遍性，一般性，多方面性，广泛性
limited	[ˈlimitid]	adj. 有限的
decade	[ˈdekeid]	n. 十年，十
introduction	[ˌintrəˈdʌkʃən]	n. 介绍，传入
evidence	[ˈevidəns]	n. 明显，显著，明白，迹象
enthusiastic	[inˌθjuːziˈæstik]	adj. 热心的，热情的
item	[ˈaitem]	n. 项目
shoplifting	[ˈʃɔpliftiŋ]	n. 入店行窃
swap	[swɔp]	v. & n. 交换
shoplifter	[ˈʃɔpliftə]	n. 商店扒手
registration	[ˌredʒisˈtreiʃən]	n. 注册，报到，登记
organization	[ˌɔːgənaiˈzeiʃən]	n. 组织，机构，团体

healthcare	[ˈhelθkɛə]	n. 医疗保健，健康护理，卫生保健
medication	[ˌmediˈkeiʃən]	n. 药物治疗，药物处理，药物
luggage	[ˈlʌgidʒ]	n. 行李，皮箱
parcel	[ˈpɑːsl]	n. 小包，包裹
		vt. 打包，捆扎
fairground	[ˈfɛəgraund]	n. 举行赛会的场所，露天市场
departure	[diˈpɑːtʃə]	n. 启程，出发，离开
holder	[ˈhəuldə]	n. 持有者，占有者，(台、架等)支持物，固定器
proprietor	[prəˈpraiətə]	n. 所有者，经营者
duplicate	[ˈdjuːplikeit]	adj. 复制的
fraudulent	[ˈfrɔːdjulənt]	adj. 欺诈的，欺骗性的
hyperlink	[ˈhaipəliŋk]	n. 超链接
cellphone	[ˈselfəun]	n. 蜂窝式便携无线电话
browse	[brauz]	v. & n. 浏览
symbology	[simˈbɔlədʒi]	n. 符号学
checksum	[tʃeksʌm]	n. 校验和
classified	[ˈklæsifaid]	v. 分类
		adj. 机密的
abut	[əˈbʌt]	v. 邻接，毗邻
vice versa	[ˈvaisi ˈvəːsə]	adv. 反之亦然
intercharacter	[ˌintəˈkæriktə]	adj. 字符间的
ignore	[igˈnɔː]	vt. 不理睬，忽视
significance	[sigˈnifikəns]	n. 意义，重要性
interleaving	[ˌintə(ː)ˈliːviŋ]	n. 交叉，交错
matrix	[ˈmeitriks]	n. 矩阵
steganography	[stəgənˈɔgrəfi]	n. 信息隐藏，速记式加密
image	[ˈimidʒ]	n. 图像
sweep	[swiːp]	v. 扫，扫过，掠过
slice	[slais]	n. 薄片，切片，一份，部分，片段
imager	[ˈimidʒə]	n. 成像器
pioneer	[ˌpaiəˈniə]	n. 先驱，倡导者，先锋
supplant	[səˈplɑːnt]	vt. 排挤掉，代替
photosensor	[ˌfəutəuˈsensə]	n. 光敏元件，光敏器件，光传感器
scrub	[skrʌb]	v. 洗擦，擦净
compatible	[kəmˈpætəbl]	adj. 兼容的
adaptor	[əˈdæptə]	n. 适配器
detailed	[ˈdiːteild]	adj. 详细的，逐条的
accelerate	[ækˈseləreit]	v. 加速，促进
confidence	[ˈkɔnfidəns]	n. 信心

merchandising	[ˈməːtʃəndaiziŋ]	n. 商品之广告推销，销售规划
predict	[priˈdikt]	v. 预知，预言，预报
seasonal	[ˈsiːzənl]	adj. 季节性的，周期性的
fluctuation	[ˌflʌktjuˈeiʃən]	n. 波动，起伏
reprice	[ˈriprais]	vt. 重新定价
reflect	[riˈflekt]	v. 反射，反映，表现
profile	[ˈprəufail]	v. 分析，介绍
voluntary	[ˈvɔləntəri]	adj. 自动的，自愿的，主动的，故意的
pitch	[pitʃ]	vt. 定位于，设定
dangerous	[ˈdeindʒərəs]	adj. 危险的
advocate	[ˈædvəkeit]	vt. 提倡，主张，拥护
logistics	[ləˈdʒistiks]	n. 物流
database	[ˈdeitəbeis]	n. 数据库，资料库
destination	[ˌdestiˈneiʃən]	n. 目的地；[计]目的文件，目的单元格
retailer	[riːˈteilə]	n. 零售商人；传播的人
shipment	[ˈʃipmənt]	n. 装船，出货
substitution	[ˌsʌbstiˈtjuːʃən]	n. 代替，取代作用，置换
error	[ˈerə]	n. 误差

✎ Phrases

be referred to	被提及；涉及
geometric pattern	几何图形
barcode reader	条码阅读机
spread to	传到，波及，蔓延到
chewing gum	口香糖
make inroads in	成功进入，进入，入侵
department store	百货公司
medical history	病史
drug allergy	药物过敏反应
conveyor line	输送线
data collection	数据收集，数据汇集，收集资料
nuclear waste	原子能工业废料
registered mail	挂号信，挂号邮件
web page	网页
in the vicinity	在附近
boarding pass	登机证
be adhered to	应坚持
in pairs	成双地，成对地

laser scanner	激光扫描仪
light beam	光束
moving part	运动机件，可动部分
custom code	自定义码
point-of-sale management	销售点管理
build up	增进，增大
error rate	差错率

Abbreviations

AIDC (Automatic Identification and Data Capture)	自动识别与数据捕捉
UPC (Universal Product Code)	通用产品码
CCD (Charge Coupled Device)	电荷耦合器件
USB (Universal Serial Bus)	通用串行总线
UID (Unique Identifying Number)	唯一识别码
EDI (Electronic Data Interchange)	电子数据交换
DC (Distribution Center)	配送中心

Exercises

I. Answer the following questions according to the text.

1. What is a barcode?
2. What and when was the very first scanning of the now ubiquitous Universal Product Code (UPC) barcode on?
3. Why does almost every item other than fresh produce from a grocery store, department store, and mass merchandiser have a UPC barcode on it?
4. What might a capable cellphone be used to do?
5. What is the mapping between messages and barcodes called?
6. What are the two properties linear symbologies can be classified mainly by?
7. What are the most common 2D symbologies?
8. Why cannot 2D symbologies be read by a laser?
9. How many categories can barcode scanners be classified into? What are they?
10. What can barcode systems provide in point-of-sale management?

II. Translate the following terms or phrases from English into Chinese and vice versa.

1. automate
2. web page
3. checksum
4. 条码阅读机

1.
2.
3.
4.

5. data collection
6. adaptor
7. steganography
8. compatible
9. n. 超链接
10. photosensor

5. _____
6. _____
7. _____
8. _____
9. _____
10. _____

III. Fill in the blanks with the words given below.

| plugged | data-entry | tested | process | downloadable |
| separate | smartphone | reinstall | scanner | identified |

❖❖❖ Guide Review-IntelliScanner Mini Portable Bar Code Scanner ❖❖❖

When I first heard about this portable bar code scanner, I couldn't quite imagine how it could be used, or why I'd need it. That changed as soon as it arrived in its shiny metal box. The __1__ itself is not much bigger than a cigarette lighter and couldn't be much simpler to use (there are only two buttons; one activates the scanning laser, the other allows you to delete an item you've just scanned).

The IntelliScanner Mini uses __2__ software rather than providing a DVD or CD. While the system requirements state it's for Windows XP or Vista, or Mac OS X 10.4 or higher, it worked fine with my Windows 7 computer. The software offers separate programs for scanning media, wine, kitchen items, home assets, and comics. Each has a __3__ activation code, and they can be purchased directly from the manufacturer's site.

Set up was fast and easy, though since I didn't do it through my Administrator log in, I ran into some issues that required me to uninstall and __4__ (so make sure you're logged on as an administrator before you start). I scanned about 90 DVDs in about five minutes, and then __5__ the reader into the computer using the included USB cable. The software reached out onto the Internet and plugged information directly into the program, a __6__ that took about six minutes. The result is a list of all the DVDs, including cast, date, price, and even a synopsis of each. Customizable fields allow you to add comments, stars, and so on. The data is as good as whatever data Amazon and other online databases hold, so you might find some odd results—for example, Bruce Willis and Madeline Stowe don't appear in the cast of 12 Monkeys, though Ernest Abuba, Bob Adrian, and Vernon Campbell, do (they were in the movie, with small roles).

I also __7__ the wine program; this one lets you scan your wine bar codes, and categorizes them in the proper areas (red, white, etc.). If you drink one, you scan it again and list it as "consumed" so that it shows up as empty—this way, next time you go shopping for wine, you know what you have and what you don't.

I had less luck with the kitchen program. Almost everything I scanned couldn't be closely __8__ —the manufacturer's name came up but not the actual product, so it was nearly impossible to know which item was which. Only big-name items (Glad Cling Wrap, for example) were properly identified. The concept is great, though; you can create shopping lists by scanning items, export your existing grocery inventory to Excel, an iPod, or a __9__, and other handy features. Finally, the home-assets program doesn't seem to reach out for information about scanned bar codes, so it's primarily a __10__ exercise; helpful for insurance inventories, however.

The scanner itself works great. It never had an issue reading a bar code, even when it was hidden below layers of clear plastic, was warped or curved, or was even partially damaged. Roughly $250 for the scanner and full software suite.

IV. Translate the following passage from English to Chinese.

✧✧ A barcode printer ✧✧

A barcode printer (or bar code printer) is a computer peripheral for printing barcode labels or tags that can be attached to physical objects. Barcode printers are commonly used to label cartons before shipment, or to label retail items with UPCs or EANs.

The most common barcode printers employ one of two different printing technologies. Direct thermal printers use a printhead to generate heat that causes a chemical reaction in specially designed paper that turns the paper black. Thermal transfer printers also use heat, but instead of reacting the paper, the heat melts a waxy or resin substance on a ribbon that runs over the label or tag material. The heat transfers ink from the ribbon to the paper. Direct thermal printers are generally less expensive, but they produce labels that can become illegible if exposed to heat, direct sunlight, or chemical vapors.

Barcode printers are designed for different markets. Industrial barcode printers are used in large warehouses and manufacturing facilities. They have large paper capacities, operate faster and have a longer service life. For retail and office environments, desktop barcode printers are most common.

How 2D Bar Codes will Work?

1. Introduction to How 2D Bar Codes Work

In the summer of 1974 at a grocery store in Troy, Ohio, an event first took place that would forever change the way we purchase things: A clerk scanned the UPC code[①] on a pack of

---

① "UPC" stands for Universal Product Code. UPC bar codes were originally created to help grocery stores speed up the checkout process and keep better track of inventory, but the system quickly spread to all other retail products because it was so successful.

Wrigley's gum. UPC codes exploded in popularity after hitting the market, and today you can find them on practically every product on the shelves. Like all bar codes, UPC codes bridged the physical and digital worlds, providing anyone with a bar code scanner instantaneous access to the data that a bar code contains.

Like all of the first bar code formats, UPC codes were 1D, meaning they only carried information in one direction. One-dimensional codes worked fine for carrying small amounts of data like numeric product codes, but as the digital world became more complex, the need for a bar code capable of carrying more data became apparent.

One solution still used today is the stacked bar code, which, as the name implies, contains a number of 1D codes piled one on top of another. Although stacked bar codes can accommodate more information than their traditional 1D counterparts, they can quickly grow very large in order to store more data and can be difficult to read. In order to have a bar code that was small in size, easy to read and capable of holding both a large amount of data and a large variety of character types, the market called for a new approach. Enter 2D bar codes(See Figure 5.2).

2D bar codes can hold mountains of information, compared to their low-tech 1D predecessors.

Figure 5.2   2D bar codes

As you might have guessed, 2D bar codes (sometimes called matrix codes) carry information in two directions: vertically and horizontally. Accordingly, 2D bar codes are capable of holding tens and even hundreds of times as much information as 1D bar codes. For instance, one of the most popular 2D bar code formats, Denso Wave's QR Code, can hold more than 7,000 digits or 4,000 characters of text, whereas even the most complex 1D codes top out around 20 characters. 2D bar codes are also small and easy to scan.

Still, 2D codes aren't perfect for every application. Because they're more complex than 1D codes, they require more powerful scanners to decode. What's more, many people are simply unfamiliar with the technology, which hinders widespread adoption. But thanks to the smartphone[①] in your pocket, that may all be about to change. Read on to find out why.

---

① A smartphone is one device that can take care of all of your handheld computing and communication needs in a single, small package.

2. 2D Bar Code Generators and Scanners

When the now ubiquitous UPC code first started making waves in the 1970s, retailers everywhere recognized the potential immediately. Unfortunately, the technology faced something of a Catch-22. Retailers refused to buy the expensive scanners needed to read the codes until manufacturers began putting UPC codes on all of their products, and manufacturers stonewalled on adopting them until they knew retailers could read the codes. Eventually, large retailers like Kmart jumped in to kick-start the technology.

Fortunately for proponents of 2D bar codes, we buy millions of scanners every year in the form of our smartphones. Common models like the iPhone, BlackBerry and Android all have the capability to read the most popular 2D bar code formats, helping to clear perhaps the largest hurdle to their widespread adoption. But how does a smartphone—or any bar code scanner, for that matter—actually make sense of the seemingly unintelligible patterns of lines and squares that 2D bar codes contain? Part of the answer lies in the design of the bar code itself, which is created from the ground up to make the scanning process as accurate and speedy as possible.

Let's check out of one of the most widespread types of 2D bar codes, QR Codes[①], to see how its design helps bar code scanners read the data it contains. For starters, every QR Code contains a finder pattern, an arrangement of squares that help the scanner detect the size of the QR Code, the direction it's facing and even the angle at which the code is being scanned. Next, every QR Code contains an alignment pattern, another pattern of squares devised to help scanners determine if the 2D bar code is distorted (perhaps it's placed on a round surface, for instance). QR Codes also have margins for error, meaning that even if part of the code is smudged or obscured, the code can often still be scanned.

But even a perfectly designed bar code would be nothing without sophisticated software capable of recognizing the bar code's alignment patterns and decoding the data(See Figure 5.3). For instance, the scanning software used to read QR Codes has some pretty impressive capabilities. Once the smartphone's camera processes the code's image, the software goes to work analyzing the image. By calculating the ratio between the black and white areas of the code, it can quickly identify which squares are part of the alignment patterns and which squares contain actual data. Using the QR Code's built-in patterns and error correction, the software can also compensate for any distortion or obscured areas of the bar code. After the software has digitally "reconstructed" the QR Code, it examines the jumble of black and white squares in the QR Code's data section and outputs the data contained within.

Of course, QR Code is only one example of a 2D bar code. For instance, the shipping

---

① A QR code (abbreviated from Quick Response code，快速响应码) is a type of matrix barcode (or two-dimensional code) first designed for the automotive industry. More recently, the system has become popular outside of the industry due to its fast readability ([ˌriːdəˈbiliti]n.易读，可读性) and comparatively ([kəmˈpærətivli]adv. 比较地，相当地) large storage capacity. The code consists of black modules arranged in a square pattern on a white background.

company UPS uses a format called MaxiCode①, which can be scanned very quickly as packages fly down the conveyor belt, whereas the U.S. Department of Defense has adopted DataMatrix②, a 2D bar code format capable of holding a lot of information in a very small area. Regardless of the format, 2D bar codes contain both data and built-in patterns to help the scanner decode the information each bar code contains, and in many cases, one device can read a variety of different formats, even traditional 1D bar codes.

Figure 5.3   Diagram of a 2D bar code

If you want to create your own, you'll find several great 2D bar code generators online. They let you adjust everything from the format you want to use to the size of the code, so the next time you're putting up a flyer for your local garage sale, consider adding a 2D bar code with your home address at the bottom. Who knows how many smartphone-carrying bargain hunters you might attract?

3. 2D Bar Code Advertising

If you can think of a way to sell a product, it's probably been done: blimps with giant company logos, televisions streaming ads to the backseats of taxicabs and even advertisements printed in edible ink on food. Nothing is off-limits from advertisers. But even they have long faced a difficult challenge as they try to determine how effective their physical ads really are; regardless of the medium, they can only guess how many people end up buying products or learning more about their company because of a particular ad. That all changed with 2D bar

---

① A MaxiCode symbol (internally called "Bird's Eye", "Target", or "ups code") appears as a 1 inch square, with a <u>bullseye</u> (靶眼，圆心) in the middle, <u>surrounded by</u> (被……环绕) a pattern of hexagonal dots. It can store about 93 characters of information, and up to 8 MaxiCode symbols can be chained together to <u>convey</u> ([kən'veɪ]vt.传达) more data. The centered symmetrical bullseye is useful in automatic symbol location regardless of orientation, and it allows MaxiCode symbols to be scanned even on a package traveling rapidly.

② A Data Matrix code is a two-dimensional matrix barcode consisting of black and white "<u>cells</u>" ([sel]n. 单元)or modules arranged in either a square or rectangular pattern. The information to be encoded can be text or raw data.

codes. For the first time, companies could simply add a 2D bar code to their advertisement and directly track how many times consumers scanned the code. If someone did scan it, the advertisers could then track whether that person went on to visit the company's Web site or even purchase a particular product.

Putting the technology to use, the fashion company Tommy Bahama added a 2D bar code into Esquire Magazine that, when scanned, took readers directly to a Web page featuring the pair of sunglasses shown in the magazine ad. With one click of their mouse (or smartphone), readers could then buy the sunglasses for a cool $138 a pop. Not only did the codes help Tommy Bahama sell a lot of sunglasses, they also helped the company learn more about their customers, telling the company what time of day and part of the country an ad was scanned.

Target is another company exploring the capabilities of 2D bar codes, adding QR codes to its magazine advertisements. Curious readers can scan the codes and instantly watch a video from a famous interior designer explaining how to use the product in the home. Advertisers say this level of interactivity is perfect for a new generation of tech-savvy consumers who want instantaneous access to product information, reviews and more. In fact, the codes have become such a popular way to advertise in Japan that they're even found on billboards, where they can be scanned at highway speeds from a passing car.

Currently, the codes aren't as popular in the United States as they are in Japan, so data on just how many advertisers are using them and what sort of results they've been getting is hard to come by. Still, as more people purchase smartphones and learn about the capabilities of QR Codes, 2D bar codes may take off much like their 1D predecessors.

New Words

gum	[gʌm]	n. 口香糖，香口胶，泡泡糖
explode	[iks'pləud]	vi. 爆发，破除，推翻，激发
popularity	[ˌpɔpju'læriti]	n. 普及，流行
practically	['præktikəli]	adv. 实际上，事实上，在实践上
		adv. 几乎，简直
instantaneous	[ˌinstən'teinjəs]	adj. 瞬间的，即刻的，即时的
format	['fɔːmæt]	n. 形式，格式
		vt. 安排……的格局(或规格)，[计]格式化(磁盘)
complex	['kɔmpleks]	adj. 复杂的，合成的，综合的
		n. 联合体
solution	[sə'ljuːʃən]	n. 解答，解决办法，解决方案
counterpart	['kauntəpɑːt]	n. 副本，极相似的人或物，配对物
guess	[ges]	v. 猜测，推测，想，认为
		n. 猜测，推测

hinder	['hində]	adj. 后面的
		v. 阻碍，打扰
generator	['dʒenəreitə]	n. 发生器
unfortunately	[ʌn'fɔːtjunətli]	adv. 不幸地
refuse	[ri'fjuːz]	vt. 拒绝，谢绝
		n. 废物，垃圾
expensive	[iks'pensiv]	adj. 花费的，昂贵的
stonewall	[stəunwɔːl]	vi. 妨碍，阻碍
kick-start	[kik-stɑːt]	vt. 发起
		n. 强劲推动力
proponent	[prə'pəunənt]	n. 建议者，支持者
blackberry	['blækbəri]	n. 黑莓
android	['ændrɔid]	n. 安卓；机器人
hurdle	['həːdl]	n. 障碍
		v. 克服(障碍)
speedy	['spiːdi]	adj. 快的，迅速的
starter	['stɑːtə]	n. 起动器
finder	['faində]	adj. 发现者，探测器
alignment	[ə'lainmənt]	n. 队列
devise	[di'vaiz]	vt. 设计，发明，图谋，做出(计划)，想出(办法)
distorted	[dis'tɔːtid]	adj. 扭歪的，受到曲解的
smudge	[smʌdʒ]	n. 污迹
		v. 弄脏，染污
ratio	['reiʃiəu]	n. 比，比率
reconstructed	[ˌriːkən'strʌktid]	adj. 重建的，改造的
jumble	['dʒʌmbl]	v. 混杂，搞乱
		n. 混乱
shipping	['ʃipiŋ]	n. 海运，运送，航行
adjust	[ə'dʒʌst]	vt. 调整，调节，校准，使适合
blimp	[blimp]	n. 软式小型飞船
logo	['lɔgəu]	n. 标识
backseat	['bæksiːt]	n. 后座，次要位置
taxicab	['tæksikæb]	n. 出租车
edible	['edibl]	adj. 可食用的
off-limit	[ɔːf-'limit]	adj. 禁止进入的
advertiser	['ædvətaizə]	n. 登广告者，广告客户
medium	['miːdjəm]	n. 媒体，方法，媒介
		adj. 中间的，中等的

interior	[in'tiəriə]	adj. 内部的，内的
		n. 内部
interactivity	[intə'æktiviti]	n. 交互，互动；人机对话
tech-savvy	[tek-'sævi]	adj. 有技术头脑的，科技通
billboard	['bilbɔ:d]	n. (户外)布告板，广告牌
		vt. 宣传

✎ Phrases

a pack of	一包，一盒
on the shelf	在货架上，在搁板上，束之高阁，不再流行，滞销的
provide … with …	为……提供……
capable of	有……能力的，可……的
Stacked Bar Code	层排式二维条码
pile on …	使堆积在……
call for	要求，提倡；为……叫喊，为……叫
top out	达到极限
unfamiliar with	不熟悉
be about to	将要，正打算
code generator	编码发生器
make waves	兴风作浪，制造纠纷
jump in	投入
for that matter	就此而言，而且，说到那一点
margin for error	误差容许量，误差限度
error correction	纠错，误差修正
Department of Defense	(美国)国防部
regardless of	不管，不顾
a variety of	多种的
put up	举起，抬起，进行，提供，表现出
garage sale	现场旧货出售
bargain hunter	买便宜货的人，投机商人
streaming ads	流广告
put … to use	开始使用
fashion company	时装公司
a pop	每件值……钱
come by	获取，得到；从旁经过，通过
take off	腾飞，突然成功

Abbreviations

UPS (United Parcel Service)　　　　　　　美国联合包裹公司

Exercises

I. Fill in the blanks with the information given in the text.

1. Like all of the first bar code formats, UPC codes were one-dimensional, meaning they only carried information _____.

2. The stacked bar code contains a number of _____.

3. Denso Wave's QR Code is one of the most popular _____, can hold more than 7,000 digits or _____, whereas even the most complex 1D codes top out around 20 characters.

4. Still, 2D codes _____ for every application. Because they're more complex than 1-D codes, they require _____.

5. iPhone, BlackBerry and Android all have the capability to _____, helping to clear perhaps the largest hurdle to their widespread adoption.

6. Even a perfectly designed bar code would be nothing without _____ capable of recognizing the bar code's alignment patterns and decoding the data.

7. Using the QR Code's built-in patterns and error correction, the software can also compensate for _____ or _____.

8. Regardless of the format, 2D bar codes contain _____ to help the scanner decode the information each bar code contains, and in many cases, one device can read a variety of different formats, even _____.

9. Putting the technology to use, the fashion company Tommy Bahama added _____ into Esquire Magazine that, when scanned, took readers directly to _____ featuring the pair of sunglasses shown in the magazine ad.

10. Currently, 2D bar codes are more popular in _____ than they are in _____.

II. Translate the following terms or phrases from English into Chinese and vice versa.

1. error correction　　　　　　1. _____
2. Stacked Bar Code　　　　　 2. _____
3. format　　　　　　　　　　3. _____
4. n. 发生器　　　　　　　　　4. _____
5. devise　　　　　　　　　　 5. _____
6. 编码发生器　　　　　　　　6. _____
7. margin for error　　　　　　7. _____

8. starter
9. alignment
10. android

8. _____
9. _____
10. _____

Text A 参考译文

条 形 码

　　条形码是光学仪器可以阅读的数据表现形式，可以用来表示附带它的物体的数据。起初，条形码通过改变平行线的宽度和间距来表示数据，也被称为线性或一维条形码(1D)。后来其发展为矩形、圆点、六边形和其他几何形状的二维条形码(2D)。虽然 2D 系统使用了多种符号，但仍然被称为条形码。原来，用叫做"条形码阅读器"的专用光学扫描仪来扫描条形码；后来，将扫描仪和解释软件用在了包括桌面打印机和智能手机的设备上。

　　条形码最早用于有轨电车的车票上，但没有取得商业上的成功，直到它们被用于自动化超市结账系统才成功，且条形码在该系统中的应用已经非常普遍。条形码(如图5.1所示)也被广泛应用于许多其他涉及自动识别和数据捕捉(AIDC)的任务。现在无处不在的通用产品代码(UPC)最早是在 1974 年 6 月用于 Wrigley 公司的口香糖包装上。

　　虽然后来其他系统也进入了 AIDC 市场，但是条形码的简单性、广泛性和低成本限制了这些系统。直到 21 世纪最初的 10 年，在条形码引入商业领域 40 年之后，像无线射频识别(RFID)这样的技术才进入商业领域。

1. 应用

　　如 UPC 这样的条形码已经成为现代文明社会无所不在的元素，全世界的商店都热衷于使用它。除生鲜外，杂货店、百货公司及超级市场中几乎全部商品都使用 UPC 条形码。这可以用来跟踪商品和防盗(包括换定价标签)，虽然扒手现在也可以打印自己的条形码。另外，零售链会员卡(主要由杂货店和如运动器材、办公品和宠物店这样特定"大包"的零售店发行)也使用条形码来唯一地识别客户，这样可以了解到消费市场及客户的个人消费模式。在购物点，通过登记的地址和电子邮件地址，购货人可以得到折扣和特价优惠。

　　通过条形码也可以组织到大量的数据。因此它们广泛使用在卫生保健站和医院，从病人识别(访问病人的数据，包括病史、麻药过敏等)到药物管理都可涉及。它们也可以用于已经扫描为图片的文件的分类和索引、跟踪生物种类的组织以及与运动称重器结合来识别运输线上物品的重量以收集数据。

　　它们也可以用来跟踪物品和人。它们可以跟踪出租车、航空行李、原子能废料、挂号信、快递邮件和包裹。条形码票允许持票人进入运动场、电影院、剧院、游乐场和乘坐交通工具，也可以用来记录出租的交通工具的出发和到达时间。这也使得识别假票更容易。在车间管理应用软件中也广泛使用条形码，雇员可以扫描工作单并跟踪完成一个工作所花费的时间。

某些二维条形码把超链接嵌入网页中。功能强大的手机也可以阅读条码和浏览所链接的网站，这可以帮助购物者找出附近某个产品的最佳定价。从 2005 年起，航空公司便在登机牌上使用 IATA 标准的二维条形码了。从 2008 年起，二维条形码可以发送到手机上作为电子登机牌。

2. 识读码制

信息和条形码之间的映射叫做识读码制。识读码制的规范包括两部分：信息的单个数字/字符的编码以及把起止标识和终止标识转换成黑条和白条的转换方式；条形码前后空间大小的确定以及校验和的计算。

线性识读码制主要可以分为以下两类：

· 连续对离散：连续识读码制的字符通常是邻接的，一个字符以白条结束，下一个字符用一个黑条开始，或者相反；离散识读码制的字符用黑条开始和结束。只要白条不是宽到可以看做代码的结束，那么字符间的白条就被忽略。

· 双宽度对多宽度：双宽度识读码制中的黑条和白条或宽或窄。宽黑条的精确宽度并不重要，只要宽黑条满足识读码制的需要即可(通常是窄黑条的 2 到 3 倍)。多宽度识读码制中的黑条和白条宽度是叫做模数的基本宽度的许多倍。大多数这样的编码使用 1、2、3、4 倍模数四种宽度。

有些识读码制使用交错式：首字符使用不同宽度的黑线编码，然后通过改变这些黑线之间白线的宽度来编码第二个字符。这样，在条形码的相同段形成成对编码。

最常用的二维识读码制是矩阵编码，其模板是按照栅格方式排列成的矩形或圆点形式。二维识读码制也可以使用圆形和其他形式，还可以使用信息隐藏法，把模块隐藏在一个图像中。

线性识读码制最适合用于激光扫描仪：光束以直线扫过条形码，并读取明暗模式中的一小段。堆栈式识读码制用于激光扫描，激光扫描仪会多次扫过条形码。

20 世纪 90 年代，Welch Allyn 倡导使用 CCD(电荷耦合器件)成像器来读条形码。成像器与激光扫描仪不一样，不需要运动部件。2007 年，线性成像器因为其耐用性已经取代了激光扫描而成为首选的扫描引擎。

二维识读码制不能用激光扫描仪读，因为激光扫描仪通常没有读取整个符号的扫描模式。它们必须使用带 CCD 或其他数码相机传感器技术的基于图像的扫描仪来扫描。

3. 扫描仪(条形码阅读器)

条形码扫描器是最早的，而且也是最便宜的扫描仪，其内置了一个固定灯管和一个光敏传感器，可手动"扫过"条形码。

按照与计算机的连接方式，条形码扫描器可以分为三类。最老式的是 RS-232 条形码扫描器，这需要专用的程序把输入数据传输给应用程序。"键盘接口扫描器"通过 PS/2 或 AT 兼容键盘适配器电缆连接到计算机，这使得条形码数据发送给计算机就像在键盘上输入数据一样。如键盘接口扫描器一样，USB 扫描器很容易安装，并且把输入数据传输给应用程序时不需要客户编码。

4. 优点

在销售点管理中，条形码系统可以提供即时更新的详细业务信息，这可加速决策并提升可信度。例如：
- 可以快速识别快销产品和自动续订货。
- 可以识别慢销产品，防止库存积压。
- 监控销售规划的执行效果，便于把那些卖得快、利润高的产品放到最好的位置。
- 可以用历史数据正确地预测季节性的销售波动。
- 可以重新对货架上的商品定价，以反映销售价格和价格增长。
- 该技术也可以通过自动注册折扣卡来收集个人客户的信息。尽管对客户来说可能会有利，但保护隐私的倡导者认为该行为存在潜在危险。

除了销售和库存跟踪外，条形码在物流领域也非常有用。
- 当工厂为出货而装箱时，可以给包装箱分配唯一的识别号(UID)。
- 一个数据库可以把 UID 与该箱的相关信息链接起来，如订单号、装箱清单、装箱数量、目的地等。
- 这些信息可以通过通信系统(如电子数据交换)传输，这样零售商在收到货物之前就知道了装运信息。
- 可以在转运之前跟踪发送给配送中心的装运货物。当货物到达目的地时，扫描 UID 就知道货物的来源、内容和成本。

条形码扫描仪成本相当低，而且比键盘输入准确得多，大约输入 15 000 到 36 万亿个字符时才会有一个差错。但准确的差错率取决于条形码的类型。

Unit 6　Radio Frequency Identification

RFID Basic

1. What is RFID?

RFID stands for Radio-Frequency IDentification. The acronym refers to small electronic devices that consist of a small chip and an antenna. The chip typically is capable of carrying 2,000 bytes of data or less.

The RFID device serves the same purpose as a bar code or a magnetic strip on the back of a credit card or ATM card; it provides a unique identifier for that object. And, just as a bar code or magnetic strip must be scanned to get the information, the RFID device must be scanned to retrieve the identifying information.

1.1　Advantages of RFID over Barcodes

A significant advantage of RFID devices over the others mentioned above is that the RFID device does not need to be positioned precisely relative to the scanner. We're all familiar with the difficulty that store checkout clerks sometimes have in making sure that a barcode can be read. And obviously, credit cards and ATM cards must be swiped through a special reader.

In contrast, RFID devices will work within a few feet (up to 20 feet for high-frequency devices) of the scanner. For example, you could just put all of your groceries or purchases in a bag, and set the bag on the scanner. It would be able to query all of the RFID devices and total your purchase immediately.

RFID technology has been available for more than fifty years. It has only been recently that the ability to manufacture the RFID devices has fallen to the point where they can be used as a "throw away" inventory or control device. Alien Technologies recently sold 500 million RFID tags to Gillette at a cost of about ten cents per tag.

One reason that it has taken so long for RFID to come into common use is the lack of standards in the industry. Most companies invested in RFID technology only use the tags to track items within their control; many of the benefits of RFID come when items are tracked from company to company or from country to country.

1.2 Common Problems with RFID

Some common problems with RFID are reader collision[①] and tag collision. Reader collision occurs when the signals from two or more readers overlap. The tag is unable to respond to simultaneous queries. Systems must be carefully set up to avoid this problem. Tag collision occurs when many tags are present in a small area; but since the read time is very fast, it is easier for vendors to develop systems that ensure that tags respond one at a time (See Figure 6.1)

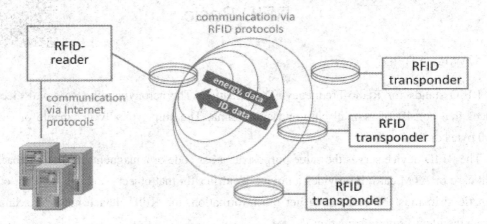

Figure 6.1　RFID communication

2. How RFID Works

How does RFID work? A Radio-Frequency IDentification system has three parts:
- A scanning antenna;
- A transceiver with a decoder to interpret the data;
- A transponder—the RFID tag—that has been programmed with information.

The scanning antenna puts out radio-frequency signals in a relatively short range. The RF radiation does two things:

[①] Reader collision occurs in RFID systems when the coverage area of one RFID reader overlaps with that of another reader. This causes two different problems: (1) Signal interference. The RF fields of two or more readers may overlap and interfere. This can be solved by having the readers programmed to read at fractionally different times. This technique (called <u>Time Division Multiple Access—TDMA</u>, 时分多址连接方式) can still result in the same tag being read twice. (2) Multiple reads of the same tag. The problem here is that the same tag is read one time by each of the overlapping readers. The only solution is to program the RFID system to make sure that a given tag (with its unique ID number) is read only once in a <u>session</u> (['seʃən]n.对话期).

• It provides a means of communicating with the transponder (the RFID tag).

• It provides the RFID tag with the energy to communicate (in the case of passive RFID tags[①]). This is an absolutely key part of the technology. RFID tags do not need to contain batteries, and can therefore remain usable for very long periods of time (maybe decades).

The scanning antennas can be permanently affixed to a surface; handheld antennas are also available. They can take whatever shape you need. For example, you could build them into a door frame to accept data from persons or objects passing through.

When an RFID tag passes through the field of the scanning antenna, it detects the activation signal from the antenna. That "wakes up" the RFID chip, and it transmits the information on its microchip to be picked up by the scanning antenna.

In addition, the RFID tag may be of one of two types. Active RFID tags[②] have their own power source; the advantage of these tags is that the reader can be much farther away and still get the signal. Even though some of these devices are built to have up to a 10 year life span, they have limited life spans. Passive RFID tags, however, do not require batteries, and can be much smaller and have a virtually unlimited life span.

RFID tags can be read in a wide variety of circumstances, where barcodes or other optically read technologies are useless.

• The tag need not be on the surface of the object (and is therefore not subject to wear);

• The read time is typically less than 100 milliseconds;

• Large numbers of tags can be read at once rather than item by item.

3. What can RFID be used for?

RFID tags come in a wide variety of shapes and sizes; they may be encased in a variety of materials:

• Animal tracking tags, inserted beneath the skin, can be rice-sized.

• Tags can be screw-shaped to identify trees or wooden items.

• Credit-card shaped for use in access applications.

• The anti-theft hard plastic tags attached to merchandise in stores are also RFID tags.

• Heavy-duty 120 by 100 by 50 millimeter rectangular transponders are used to track shipping containers, or heavy machinery, trucks, and railroad cars.

RFID devices have been used for years to identify dogs, for a means of permanent identification. Dog owners had long used tattoos, permanent ink markings, typically on the ears.

---

① A passive tag is an RFID tag that does not contain a battery; the power is supplied by the reader. When radio waves from the reader are encountered by a passive RFID tag, the coiled ([kɔil]v. 盘绕，卷) antenna within the tag forms a magnetic field (磁场). The tag draws power from it, energizing the circuits in the tag. The tag then sends the information encoded in the tag's memory.

② An RFID tag is an active tag when it is equipped with a battery that can be used as a partial or complete source of power for the tag's circuitry and antenna. Some active tags contain replaceable batteries for years of use; others are sealed ([siːld]adj. 密封的) units.

However, these can fade with age and it may be difficult to get the animal to sit still while you examine him for markings.

Many musical instruments are stolen every year. For example, custom-built or vintage guitars are worth as much as $50,000 each. Snagg, a California company specializing in RFID microchips for instruments, has embedded tiny chips in 30,000 Fender guitars already. The database of RFID chip IDs is made available to law enforcement officials, dealers, repair shops and luthiers.

4. Is RFID Technology Secure and Private?

Unfortunately, not very often in the systems to which consumers are likely to be exposed. Anyone with an appropriately equipped scanner and close access to the RFID device can activate it and read its contents. Obviously, some concerns are greater than others. If someone walks by your bag of books from the bookstore with a 13.56 MHz "sniffer" with an RF field that will activate the RFID devices in the books you bought, that person can get a complete list of what you just bought. That's certainly an invasion of your privacy, but it could be worse. Another scenario involves a military situation in which the other side scans vehicles going by, looking for tags that are associated with items that only high-ranking officers can have, and targeting accordingly.

Companies are more concerned with the increasing use of RFID devices in company badges. An appropriate RF field will cause the RFID chip in the badge to "spill the beans" to whomever activates it. This information can then be stored and replayed to company scanners, allowing the thief access—and your badge is the one that is "credited" with the access.

The smallest tags that will likely be used for consumer items don't have enough computing power to do data encryption to protect your privacy. The most they can do is PIN-style or password-based protection.

5. Next-Generation Uses of RFID?

Some vendors have been combining RFID tags with sensors of different kinds. This would allow the tag to report not simply the same information over and over, but identifying information along with current data picked up by the sensor. For example, an RFID tag attached to a leg of lamb could report on the temperature readings of the past 24 hours, to ensure that the meat was properly kept cool.

Over time, the proportion of "scan-it-yourself" aisles in retail stores will increase. Eventually, we may wind up with stores that have mostly "scan-it-yourself" aisles and only a few checkout stations for people who are disabled or unwilling.

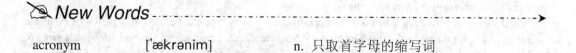

New Words

acronym	['ækrənim]	n. 只取首字母的缩写词

chip	[tʃip]	n.	芯片
antenna	[æn'tenə]	n.	天线
position	[pə'ziʃən]	vt.	安置，决定……的位置
precisely	[pri'saisli]	adv.	正好
swipe	[swaip]	vt.	刷……卡
query	['kwiəri]	n.	质问，询问，怀疑，疑问
		v.	询问
collision	[kə'liʒən]	n.	碰撞，冲突
simultaneous	[ˌsiməl'teinjəs]	adj.	同时的，同时发生的
vendor	['vendɔ:]	n.	卖主
ensure	[in'ʃuə]	v.	确保，保证
transponder	[træn'spɔndə]	n.	变换器，转调器；异频雷达收发机
energy	['enədʒi]	n.	精力，活力；能量
absolutely	['æbsəlu:tli]	adv.	完全地，绝对地
permanently	['pə:mənəntli]	adv.	永存地，不变地
handheld	['hænd,held]	adj.	掌上型，手持型
shape	[ʃeip]	n.	外形，形状
		vi.	成形
microchip	['maikrəutʃip]	n.	微芯片
circumstance	['sə:kəmstəns]	n.	环境，详情，境况
encase	[in'keis]	vt.	包括，装入，包住，围
insert	[in'sə:t]	vt.	插入，嵌入
		n.	插入物
rice-sized	[rais-saizd]	adj.	米粒大小的
anti-theft	['ænti-θeft]	adj.	防盗的
merchandise	['mə:tʃəndaiz]	n.	商品，货物
millimeter	['milimi:tə]	n.	毫米
rectangular	[rek'tæŋgjulə]	adj.	矩形的，成直角的
container	[kən'teinə]	n.	容器，集装箱
tattoo	[tə'tu:]	n.	纹身
fade	[feid]	vi.	减弱下去，消失
custom-built	['kʌstəm'bilt]	adj.	定制的，定做的
vintage	['vintidʒ]	adj.	古老的，最佳的，过时的
dealer	['di:lə]	n.	经销商，商人
luthier	['lu:tiə]	n.	拨弦乐器制作匠
sniffer	['snifə]	n.	嗅探器
invasion	[in'veiʒən]	n.	入侵
badge	[bædʒ]	n.	徽章，证章
aisle	[ail]	n.	走廊，过道

Phrases

serve the same purpose as	与……目的一样
magnetic strip	磁条，磁片
credit card	信用卡
relative to	相对于
in contrast	相反，大不相同
high-frequency device	高频设备
invest in	投资于，买进
respond to …	对……做出响应
in the case of …	在……的情况
passive RFID tag	无源 RFID 标签
be affixed to	被贴到
door frame	门框
wake up	醒来，唤醒
life span	寿命，使用期限
active RFID tag	有源 RFID 标签
with age	随着年龄的增长，因年久
law enforcement official	执法人员
repair shop	维修车间
be associated with	和……联系在一起；与……有关
spill the beans	泄露秘密，说漏嘴
over and over	反复，再三
a leg of lamb	羊腿
wind up with …	以……结束
checkout station	结账台

Abbreviations

ATM (Automatic Teller Machine)   自动取款（出纳）机
PIN (Personal Identification Number)   个人身份号码

Exercises

I. Answer the following questions according to the text.

1. What does RFID stand for? What does it refer to?
2. What is the significant advantage of RFID devices over the others?
3. What is one reason that it has taken so long for RFID to come into common use?
4. What are some common problems with RFID?

5. How many parts does a Radio-Frequency IDentification system have? What are they?
6. What does the RF radiation do?
7. What are the two types of the RFID tags?
8. What are the circumstances in which RFID tags can be read?
9. What are companies more concerned with?
10. What may we wind up with eventually?

II. Translate the following terms or phrases from English into Chinese and vice versa.

1. chip
2. collision
3. n. 嗅探器
4. passive RFID tag
5. antenna
6. transponder
7. high-frequency device
8. active RFID tag
9. 磁条，磁片
10. microchip

1. _____
2. _____
3. _____
4. _____
5. _____
6. _____
7. _____
8. _____
9. _____
10. _____

III. Fill in the blanks with the words given below.

| customers | experimenting | maintenance | advantage | logistics |
| track | electronic | assets | payment | control |

◆◆◆ Current and Potential Uses of RFID ◆◆◆

1. Asset Tracking

It's no surprise that asset tracking is one of the most common uses of RFID. Companies can put RFID tags on ___1___ that are lost or stolen often, that are underutilized or that are just hard to locate at the time they are needed. Just about every type of RFID system is used for asset management. NYK Logistics, a third-party ___2___ provider based in Secaucus, N.J., needed to track containers at its Long Beach, Calif., distribution center. It chose a real-time locating system that uses active RFID beacons to locate container to within 10 feet.

2. Manufacturing

RFID has been used in manufacturing plants for more than a decade. It's used to ___3___ parts and work in process and to reduce defects, increase throughput and manage the production of different versions of the same product.

3. Supply Chain Management

RFID technology has been used in closed loop supply chains or to automate parts of the supply chain within a company's ___4___ for years.

As standards emerge, companies are increasingly turning to RFID to track shipments among supply chain partners.

4. Retailing

Retailers such as Best Buy, Metro, Target, Tesco and Wal-Mart are in the forefront of RFID adoption. These retailers are currently focused on improving supply chain efficiency and making sure product is on the shelf when ___5___ want to buy it.

5. Payment Systems

RFID is all the rage in the supply chain world, but the technology is also catching on as a convenient ___6___ mechanism. One of the most popular uses of RFID today is to pay for road tolls without stopping. These active systems have caught on in many countries, and quick service restaurants are ___7___ with using the same active RFID tags to pay for meals at drive-through windows.

6. Security and Access Control

RFID has long been used as an ___8___ key to control who has access to office buildings or areas within office buildings. The first access control systems used low-frequency RFID tags. Recently, vendors have introduced 13.56 MHz systems that offer longer read range. The ___9___ of RFID is it is convenient (an employee can hold up a badge to unlock a door, rather than looking for a key or swiping a magnetic stripe card) and because there is no contact between the card and reader, there is less wear and tear, and therefore less ___10___.

As RFID technology evolves and becomes less expensive and more robust, it's likely that companies and RFID vendors will develop many new applications to solve common and unique business problems.

IV. Translate the following passage from English to Chinese.

✦✦ Components of Radio Frequency IDentification (RFID) ✦✦

A basic RFID system consists of three components:
- An antenna or coil;
- A transceiver (with decoder);
- A transponder (RF tag) electronically programmed with unique information.

The antenna emits radio signals to activate the tag and to read and write data to it.

The reader emits radio waves in ranges of anywhere from one inch to 100 feet or more, depending upon its power output and the radio frequency used. When an RFID tag passes through the electromagnetic zone, it detects the reader's activation signal.

The reader decodes the data encoded in the tag's integrated circuit (silicon chip) and the data is passed to the host computer for processing.

The purpose of an RFID system is to enable data to be transmitted by a portable device, called a tag, which is read by an RFID reader and processed according to the needs of a particular application. The data transmitted by the tag may provide identification or location information, or specifics about the product tagged, such as price, color, date of purchase, etc. RFID technology has been used by thousands of companies for a decade or more. RFID quickly gained attention because of its ability to track moving objects. As the technology is refined, more pervasive uses for RFID tags are in the works.

A typical RFID tag consists of a microchip attached to a radio antenna mounted on a substrate. The chip can store as much as 2 kilobytes of data.

To retrieve the data stored on an RFID tag, you need a reader. A typical reader is a device that has one or more antennas that emit radio waves and receive signals back from the tag. The reader then passes the information in digital form to a computer system.

How RFID Works?

1. Introduction to How RFID Works

Long checkout lines at the grocery store are one of the biggest complaints about the shopping experience. Soon, these lines could disappear when the ubiquitous Universal Product Code (UPC) bar code is replaced by smart labels, also called Radio Frequency IDentification (RFID) tags. RFID tags are intelligent bar codes that can talk to a networked system to track every product that you put in your shopping cart.

Imagine going to the grocery store, filling up your cart and walking right out of the door. No longer will you have to wait as someone rings up each item in your cart one at a time. Instead, these RFID tags will communicate with an electronic reader that will detect every item in the cart and ring each up almost instantly. The reader will be connected to a large network that will send information on your products to the retailer and product manufacturers. Your bank will then be notified and the amount of the bill will be deducted from your account. No lines, no waiting.

RFID tags, a technology once limited to tracking cattle, are tracking consumer products worldwide. Many manufacturers use the tags to track the location of each product they make from the time it's made until it's pulled off the shelf and tossed in a shopping cart.

Outside the realm of retail merchandise, RFID tags are tracking vehicles, airline passengers, Alzheimer's patients and pets. Soon, they may even track your preference for chunky or creamy peanut butter. Some critics say RFID technology is becoming too much a part of our lives—that is, if we're even aware of all the parts of our lives that it affects.

2. Reinventing the Bar Code

Almost everything that you buy from retailers has a UPC bar code printed on it. These bar codes help manufacturers and retailers keep track of inventory. They also give valuable information about the quantity of products being bought and, to some extent, the consumers buying them. These codes serve as product fingerprints① made of machine-readable parallel bars that store binary code.

Created in the early 1970s to speed up the check out process, bar codes have a few disadvantages:

• In order to keep up with inventories, companies must scan each bar code on every box of a particular product.

• Going through the checkout line involves the same process of scanning each bar code on each item.

• Bar code is a read-only technology, meaning that it cannot send out any information.

RFID tags are an improvement over bar codes because the tags have read and write capabilities. Data stored on RFID tags can be changed, updated and locked. Some stores that have begun using RFID tags have found that the technology offers a better way to track merchandise for stocking and marketing purposes. Through RFID tags, stores can see how quickly the products leave the shelves and which shoppers are buying them.

RFID tags won't entirely replace bar codes in the near future—far too many retail outlets currently use UPC scanners in billions of transactions every year. But as time goes on we'll definitely see more products tagged with RFIDs and an increased focus on seamless wireless transactions like that rosy instant checkout picture painted in the introduction. In fact, the world is already moving toward using RFID technology in payments through special credit cards and smart phones—we'll get into that later.

In addition to retail merchandise, RFID tags have also been added to transportation devices like highway toll passcards and subway passes. Because of their ability to store data so efficiently, RFID tags can tabulate the cost of tolls and fares and deduct the cost electronically from the amount of money that the user places on the card. Rather than waiting to pay a toll at a tollbooth or shelling out coins at a token counter, passengers use RFID chip-embedded passes like debit cards.

But would you entrust your medical history to an RFID tag? How about your home address or your baby's safety? Let's look at two types of RFID tags and how they store and transmit data

---

① Fingerprints are the tiny ridges, <u>whorls</u> ([hwə:l]n. 螺旋环, 涡) and valley patterns on the tip of each finger. They form from pressure on a baby's tiny, developing fingers in the womb. No two people have been found to have the same fingerprints—they are totally unique. There's a one in 64 billion chance that your fingerprint will match up exactly with someone else's.

before we move past grocery store purchase s to human lives.

3. RFID Tags Past and Present

RFID technology has been around since 1970, but until recently, it has been too expensive to use on a large scale. Originally, RFID tags were used to track large items, like cows, railroad cars and airline luggage, which were shipped over long distances. These original tags, called inductively coupled RFID tags, were complex systems of metal coils, antennae and glass.

Inductively coupled RFID tags were powered by a magnetic field generated by the RFID reader. Electrical current has an electrical component and a magnetic component—it is electromagnetic. Because of this, you can create a magnetic field with electricity, and you can create electrical current with a magnetic field. The name "inductively coupled" comes from this process—the magnetic field inducts a current in the wire (See Figure 6.2).

RFID tags like these used to be made only for tracking luggage and large parcels.

Figure 6.2　RFID tags

Capacitively coupled tags were created next in an attempt to lower the technology's cost. These were meant to be disposable tags that could be applied to less expensive merchandise and made as universal as bar codes. Capacitively coupled tags used conductive carbon ink instead of metal coils to transmit data. The ink was printed on paper labels and scanned by readers. Motorola's BiStatix RFID tags were the frontrunners in this technology. They used a silicon chip that was only 3 millimeters wide and stored 96 bits of information. This technology didn't catch on with retailers, and BiStatix was shut down in 2001.

Newer innovations in the RFID industry include active, semi-active and passive RFID tags. These tags can store up to 2 kilobytes of data and are composed of a microchip, antenna and, in the case of active and semi-passive tags, a battery. The tag's components are enclosed within plastic, silicon or sometimes glass.

At a basic level, each tag works in the same way:
- Data stored within an RFID tag's microchip waits to be read.
- The tag's antenna receives electromagnetic energy from an RFID reader's antenna.
- Using power from its internal battery or power harvested from the reader's electromagnetic field, the tag sends radio waves back to the reader.

• The reader picks up the tag's radio waves and interprets the frequencies as meaningful data.

Inductively coupled and capacitively coupled RFID tags aren't used as commonly today because they are expensive and bulky. In the next section, we'll learn more about active, semi-passive and passive RFID tags.

4. Active, Semi-passive and Passive RFID Tags

Active, semi-passive and passive RFID tags are making RFID technology more accessible and prominent in our world. These tags are less expensive to produce, and they can be made small enough to fit on almost any product.

Active and semi-passive RFID tags use internal batteries to power their circuits. An active tag also uses its battery to broadcast radio waves to a reader, whereas a semi-passive tag relies on the reader to supply its power for broadcasting. Because these tags contain more hardware than passive RFID tags, they are more expensive. Active and semi-passive tags are reserved for costly items that are read over greater distances—they broadcast high frequencies from 850 to 950 MHz that can be read 100 feet (30.5 meters) or more away. If it is necessary to read the tags from even farther away, additional batteries can boost a tag's range to over 300 feet (100 meters).

Like other wireless devices, RFID tags broadcast over a portion of the electromagnetic spectrum. The exact frequency is variable and can be chosen to avoid interference with other electronics or among RFID tags and readers in the form of tag interference or reader interference. RFID systems can use a cellular system called Time Division Multiple Access (TDMA) to make sure the wireless communication is handled properly.

Passive RFID tags rely entirely on the reader as their power source. These tags are read up to 20 feet (six meters) away, and they have lower production costs, meaning that they can be applied to less expensive merchandise. These tags are manufactured to be disposable, along with the disposable consumer goods on which they are placed. Whereas a railway car would have an active RFID tag, a bottle of shampoo would have a passive tag.

Another factor that influences the cost of RFID tags is data storage. There are three storage types: read-write, read-only and WORM (Write Once, Read Many). A read-write tag's data can be added to or overwritten. Read-only tags cannot be added to or overwritten—they contain only the data that is stored in them when they were made. WORM tags can have additional data (like another serial number) added once, but they cannot be overwritten.

Most passive RFID tags cost between seven and 20 cents U.S. each. Active and semi-passive tags are more expensive, and RFID manufacturers typically do not quote prices for these tags without first determining their range, storage type and quantity. The RFID industry's goal is to get the cost of a passive RFID tag down to five cents each once more merchandisers adopt it.

5. Talking Tags

When the RFID industry is able to lower the price of tags, it will lead to a ubiquitous network of smart packages that track every phase of the supply chain. Store-shelves will be full of smart-labeled products that can be tracked from purchase to trash can. The shelves themselves will communicate wirelessly with the network. The tags will be just one component of this large product-tracking network.

The other two pieces to this network will be the readers that communicate with the tags and the Internet, which will provide communications lines for the network.

Let's look at a real-world scenario of this system:

• At the grocery store, you buy a carton of milk. The milk containers will have an RFID tag that stores the milk's expiration date and price. When you lift the milk from the shelf, the shelf may display the milk's specific expiration date, or the information could be wirelessly sent to your personal digital assistant or cell phone.

• As you exit the store, you pass through doors with an embedded tag reader. This reader tabulates the cost of all the items in your shopping cart and sends the grocery bill to your bank, which deducts the amount from your account. Product manufacturers know that you've bought their product, and the store's computers know exactly how many of each product need to be reordered.

• Once you get home, you put your milk in the refrigerator, which is also equipped with a tag reader. This smart refrigerator is capable of tracking all of the groceries stored in it. It can track the foods you use and how often you restock your refrigerator, and can let you know when that milk and other foods spoil.

• Products are also tracked when they are thrown into a trash can or recycle bin. At this point, your refrigerator could add milk to your grocery list, or you could program the fridge to order these items automatically.

• Based on the products you buy, your grocery store gets to know your unique preferences. Instead of receiving generic newsletters with weekly grocery specials, you might receive one created just for you. If you have two school-age children and a puppy, your grocery store can use customer-specific marketing by sending you coupons for items like juice boxes and dog food.

In order for this system to work, each product will be given a unique product number. MIT's Auto-ID Center is working on an Electronic Product Code (EPC) identifier that could replace the UPC. Every smart label could contain 96 bits of information, including the product manufacturer, product name and a 40-bit serial number. Using this system, a smart label would communicate with a network called the Object Naming Service. This database would retrieve information about a product and then direct information to the manufacturer's computers.

The information stored on the smart labels would be written in a Product Markup Language (PML), which is based on the eXtensible Markup Language (XML)[①]. PML would allow all

---

[①] Extensible Markup Language (XML) is a markup language that defines a set of rules for encoding documents in a format that is both <u>human-readable</u> (人可读) and <u>machine-readable</u> (机器可读). The design goals of XML emphasize simplicity, generality, and usability over the Internet. It is a textual data format with strong support via <u>Unicode</u> (统一字符码) for the languages of the world. Although the design of XML focuses on documents, it is widely used for the representation of arbitrary data structures, for example in web services.

computers to communicate with any computer system similar to the way that Web servers read HyperText Markup Language (HTML)①, the common language used to create Web pages.

We're not at this point yet, but RFID tags are more prominent in your life than you may realize. Wal-Mart and Best Buy are just two major merchandisers that use RFID tags for stocking and marketing purposes. Automated systems called intelligent software agents manage all the data coming in and going out from RFID tags and will carry out a specific course of action like sorting items.

The United States retail market is on the cusp of embracing a major implementation of RFID technology through payment systems that use Near Field Communication②. These are the credit cards of the future.

6. Near Field Communication, Smart Phones and RFID

NFC technology is promising because it presents the next evolution of convenient payment with an added layer of security. Some credit cards have NFC chips embedded in them and can be tapped against NFC payment terminals instead of swiped, which eliminates the possibility that someone could skim your data via the magnetic strip. This same system works with cellular phones, too: read up on how cellular electronic payments work to dig into the technology.

Google is one company pushing NFC payments with Google Wallet. The application stores credit card information under multiple layers of security and allows for quick tap payments at NFC terminals. That means the technology's usefulness is limited by the number of NFC payment terminals available in retail locations and the number of phones that support the technology—at launch, Google Wallet only works with the Android Nexus S smart phone.

So what does this have to do with RFID? Near Field Communication devices can read passive RFID tags and extract the information stored in them. This technology is being used in modern advertising. For example, picture a normal poster advertising a pair of jeans, the kind of paper you'd see plastered on a wall in a shopping mall. Advertisers can make "smart" posters with RFID tags that add a new level of interaction with customers. Tap an NFC phone against a "smart" poster equipped with an RFID tag, and you may get a 10 percent off coupon for those jeans at Macy's. Passive RFID tags are cheap enough to be used in promotional materials just to engage customers.

NFC and RFID technologies have huge futures ahead of them in the retail world, but security remains a common concern. Some critics find the idea of merchandisers tracking and recording purchases to be alarming.

---

① HyperText Markup Language (HTML) is the predominant markup language for web pages. HTML elements are the basic building-blocks of web pages.

② Near Field Communication (NFC) is a set of standards for smartphones and similar devices to establish radio communication with each other by touching them together or bringing them into close proximity, usually no more than a few centimeters. Present and anticipated applications include <u>contactless</u> (['kɔntæktlis]adj. 不接触的，遥控的) transactions, data exchange, and simplified setup of more complex communications such as WiFi. Communication is also possible between an NFC device and an <u>unpowered</u> ([ʌn'pauəd]adj. 无动力的，无电源的) NFC chip, called a "tag".

New Words

complaint	[kəm'pleint]	n.	投诉；诉苦，抱怨，牢骚
imagine	[i'mædʒin]	vt.	想象，设想
deduct	[di'dʌkt]	vt.	扣除
cattle	['kætl]	n.	牛，家养牲畜
toss	[tɔs]	v.	投，掷
realm	[relm]	n.	领域
chunky	['tʃʌŋki]	adj.	粗短的，厚实的
reinvent	[,ri:in'vent]	vt.	彻底改造，重新使用
quantity	['kwɔntiti]	n.	量，数量
fingerprint	['fiŋgəprint]	n.	指纹，手印
		vt.	采指纹
disadvantage	[,disəd'vɑ:ntidʒ]	n.	不利条件，缺点，劣势
outlet	['autlet]	n.	出口，出路
definitely	['definitli]	adv.	明确地，干脆地
rosy	['rəuzi]	adj.	蔷薇色的，玫瑰红色的
tabulate	['tæbjuleit]	v.	把……制成表格，以列表形式排列，列表
tollbooth	['təulbu:θ]	n.	过路收费亭
inductively	[in'dʌktivli]	adv.	感应地，诱导地，归纳地
coil	[kɔil]	v.	盘绕，卷
electromagnetic	[ilektrəu'mægnitik]	adj.	电磁的
induct	[in'dʌkt]	v.	感应
frontrunner	['frʌntrʌnə]	n.	领跑者，前锋
bulky	['bʌlki]	adj.	大的，容量大的，体积大的
spectrum	['spektrəm]	n.	光谱，频谱
shelf	[ʃelf]	n.	架子，货架，书架
exit	['eksit]	n.	出口，退场
		vi.	退出，脱离
restock	['ri:'stɔk]	vt.	重新进货，再储存
newsletter	['nju:z,letə]	n.	时事通讯
grocery	['grəusəri]	n.	[美]食品杂货店，食品，杂货
puppy	['pʌpi]	n.	小狗，小动物
skim	[skim]	v.	快读，浏览；撇去
tap	[tæp]	v.	轻打，轻敲
plaster	['plɑ:stə]	vt.	在……上大量粘贴；贴满
engage	[in'geidʒ]	vt.	吸引

📝 Phrases

be replaced by	被代替
shopping cart	购物车
fill up	填补，装满
one at a time	每次一个
deduct from	扣除
pull off	努力实现，赢得
peanut butter	花生酱
be aware of	知道
to some extent	在某种程度上，(多少)有一点
keep up with	跟上
instant checkout picture	即时检测图像
paint in	补画上去，加绘上去
highway toll passcard	高速公路收费卡
subway pass	地铁通行卡
pay a toll	付通行费
shell out	交付，支付
token counter	收银台，柜台
debit card	借记卡
on a large scale	大规模地
inductively coupled RFID tag	电感耦合 RFID 标签
magnetic field	磁场
electrical current	电流
capacitively coupled tag	电容耦合标签
in an attempt to	力图，试图
conductive carbon ink	导电碳墨
silicon chip	硅片
catch on with	迎合，跟随
shut down	放下，关下，(使)机器等关闭，停车
trash can	垃圾桶，垃圾箱
expiration date	产品有效期
recycle bin	回收站
serial number	序列号，系列号
Object Naming Service	物件名称解析服务
on the cusp of	正在着手，正处于
plaster on	在……上面涂抹或粘贴
shopping mall	大型购物中心

Abbreviations

TDMA (Time Division Multiple Access)　　时分多址访问
WORM (Write Once, Read Many)　　单次写多次读
PML (Product Markup Language)　　产品标识语言
HTML (Hyper Text Markup Language)　　超文本标识语言

Exercises

I. Fill in the blanks with the information given in the text.

1. RFID tags are _____ that can talk to a networked system to track every product that you put in your shopping cart.

2. One of the disadvantages of bar codes is that companies _____ on every box of a particular product in order to keep up with inventories.

3. Data stored on RFID tags can be _____, _____ and _____.

4. _____, RFID tags can tabulate the cost of tolls and fares and deduct the cost electronically from the amount of money that the user places on the card.

5. Newer innovations in the RFID industry include _____, _____ and _____. These tags can store up to _____ of data.

6. Active and semi-passive RFID tags use _____ to power their circuits. Passive RFID tags _____ as their power source.

7. Another factor that influences the cost of RFID tags is _____. There are three storage types: _____, _____ and _____.

8. The information stored on the smart labels would be written in _____, which is based on the eXtensible Markup Language (XML).

9. Two major merchandisers that use RFID tags for stocking and marketing purposes are _____ and _____.

10. NFC technology is promising because it presents the next evolution of _____ with an added layer of security.

II. Translate the following terms or phrases from English into Chinese and vice versa.

1. spectrum　　　　　　　　　　1. _____
2. adj. 电磁的　　　　　　　　　2. _____
3. Near Field Communication　　 3. _____
4. silicon chip　　　　　　　　　4. _____
5. inductively coupled RFID tag　 5. _____
6. 序列号，系列号　　　　　　　6. _____
7. inductively　　　　　　　　　7. _____

8. capacitively coupled tag   8. _____
9. magnetic field   9. _____
10. conductive carbon ink   10. _____

Text A 参考译文

RFID 基础

1. 什么是 RFID？

RFID 译为射频识别，该缩写对应于包含一个小芯片和一个小天线的小型电子设备。其中的芯片通常能存储 2000 字节的数据。

RFID 设备的作用与条形码或银行卡或 ATM 卡背面的磁条一样，它提供物体的唯一性标识。并且，正如条形码和磁条必须通过扫描来获得信息一样，RFID 设备也必须通过扫描来检索标识信息。

1.1 RFID 比条形码更佳

与条形码和磁条相比，RFID 的一个重要优点是它不需要放置到正对扫描仪的位置就可获取信息。我们都知道商店中的员工要确保条形码的读取有多么困难。并且众所周知，信用卡和 ATM 卡必须用特制的设备来刷。

相反，RFID 设备可以在距扫描仪数英尺的范围之内工作(高频设备的工作范围可以高达 20 英尺)。例如，你可以把全部食品或购买物装到包中，然后把包放到扫描仪上，此时扫描仪能查询全部的 RFID 设备并立即统计你的购买物。

虽然 RFID 技术已经用了 50 多年，但是直到最近，制造 RFID 设备的能力才发展到允许它们可以用作"可抛弃"库存清单和控制装置的水平。Alien Technologies 公司最近以每个标签大约 10 分钱的价格向 Gillette 公司出售了 5 亿个 RFID 标签。

RFID 用了这么长的时间才进入到普通应用的原因之一是缺乏行业标准。大多数投资 RFID 技术的公司只是利用对标签的控制来跟踪项目。当跟踪的项目在公司间或者国家间流动时，RFID 的许多优势就显现出来了。

1.2 RFID 的常见问题

RFID 的一些常见问题是阅读器冲突或标签冲突。当来自两个或多个阅读器的信号重叠时，就会发生阅读器冲突。由于标签不能响应并发查询，因此必须小心地设置系统以避免这个问题。当多个标签出现在一个小区域时，则会发生标签冲突。但因为读的速度非常快，经销商就更容易开发出确保标签一次只响应一个的系统(如图 6.1 所示，图略)。

2. RFID 如何工作

RFID 是如何工作的？射频识别系统有三个部分：
- 一个扫描天线；
- 一个带有解释数据的解码器的收发器；

● 一个变换器——RFID 标签,带有编程信息。

扫描天线会在一个相对小的范围发射无线电射频信号。无线电射线完成以下两个工作:
● 它提供扫描仪与变换器(RFID 标签)的通信方式。
● 它给 RFID 提供通信能量(在被动标签的情况下)。这绝对是扫描天线技术的关键部分。这样的话,RFID 标签无需包含电池,并且可以保持很长有效期(也许数十年)。

扫描天线可以永久地贴在一个平面上,手持也可以,也可以做成所需的任何形状。例如,扫描天线可以内置在一个门框内来接收经过的人或物品的数据。

当一个 RFID 标签经过扫描天线的区域时,它将探测到来自该天线的激活信号。这就"唤醒"了 RFID 芯片,并使其将微芯片上的信息发送给扫描天线。

另外,RFID 标签有两种类型可供选择:有源 RFID 标签和无源 RFID 标签。有源 RFID 标签有自己的电源,优点是即使阅读器放在很远的地方,仍然可以读取到信号。尽管这些设备中有些的寿命可以高达 10 年,但其寿命也是有限的。与之相反,无源 RFID 标签不需要电池,所以体积可以小得多并且实际上寿命不受限制。

RFID 标签在许多条形码或其他光阅读技术都不可用的环境下可被读取。原因如下:
● RFID 标签不需要放在物体表面(因此不会磨损);
● 读取的时间通常少于 100 ms;
● 大量的标签可以同时读取而不是逐项读取。

3. RFID 可以用来做什么?

RFID 标签的外形和大小各异,且它们可以封装在多种材料中:
● 动物跟踪标签可插入到皮肤下面,有米粒大小。
● 标签可以是螺丝形状的,以识别树或木制品。
● 标签可做成信用卡形状以用于访问应用程序。
● 商店中贴在商品上的防窃硬塑料标签也是 RFID 标签。
● 重型的 120 mm × 100 mm × 50 mm 的矩形转发器可以用来跟踪集装箱或重型机械、卡车及火车车厢。

多年来,RFID 一直用于识别狗,并且将作为一个永久识别的方法。长久以来,狗主人通过在狗耳朵上刻花纹、永固墨水记号来识别它。但是,时间长了这些记号都可能消失,且在检查狗身上的记号时,要让狗蹲着不动可能是困难的。

每年都有许多乐器丢失。例如,一个客户定制的或古老的吉他可能会因价值超过 50 000 美元而被盗窃。一家加利福尼亚州的公司 Snagg 专门为这些乐器研制了一种 RFID 微芯片,且已经在 30 000 个 Fender 吉他中嵌入了小芯片,同时,提供给执法人员、经销商、维修车间和制琴师的 RFID 芯片 IDs 数据库已经完成。

4. RFID 技术是安全和保密的吗?

不幸的是,RFID 设备并不是只开放给指定的用户,而是任何装备了适当扫描仪并可以访问 RFID 设备的人都可以激活它并读取其内容。明显地,有些担心更甚:如果一个人带有 13.56 MHz 的 RF 场"嗅探器",那他从你的书包旁走过时,将激活你所购买的书中的 RFID 设备,他就可以得到你刚才在书店购书的全部清单。这绝对侵犯了你的隐私,但还有比这更糟糕的。假想涉及军事的情景,敌人扫描从旁边经过的车辆,然后专门寻找高级军官携带的物品的相关标签并加以跟踪。

公司将在公司徽章中越来越多地使用RFID设备。但一个适当的RF场会引起徽章中的RFID芯片"泄密"给激活它的任何人,即允许窃贼进门——你的徽章就是可以访问的"通行证"。

最小的用于消费者项目上的标签没有足够的计算能力来加密数据以保护隐私。它们最多就是使用PIN式或输入密码来保护隐私。

5. RFID未来的用途是什么?

一些经销商已经把RFID标签与传感器进行了多种组合。这样就允许标签不只是简单地一遍遍重复报告相同的信息,而是随同该传感器获取的当前数据一起识别信息。例如,一个带在小羊腿上的RFID标签可以报告过去24小时它所读取的温度,以确保肉食保持适当的低温。

随着时间的流逝,商店中"自助扫描"的通道比例会不断增加。不久之后,我们也许会看到商店主要使用"自助扫描"通道,而只为伤残或不愿使用此通道的人提供很少的结账台。

Unit 7　WiFi and Bluetooth

How WiFi Works?

If you've been at an airport, in a coffee shop, a library or a hotel recently, chances are you've been right in the middle of a wireless network. Many people also use wireless networking, also called WiFi or 802.11 networking, to connect their computers at home, and some cities are trying to use the technology to provide free or low-cost Internet access to residents. In the near future, wireless networking may become so widespread that you can access the Internet just about anywhere at any time, without using wires (See Figure 7.1).

Wireless networks make it easy to connect to the Internet.
Figure 7.1　WiFi works without using wires

WiFi has a lot of advantages. Wireless networks are easy to set up and inexpensive. They're also unobtrusive—unless you're on the lookout for a place to use your laptop, you may not even notice when you're in a hotspot. In this article, we'll look at the technology that allows information to travel over the air. We'll also review what it takes to create a wireless network in your home.

1. What Is WiFi?

A wireless network uses radio waves, just like cell phones, televisions and radios do. In fact, communication across a wireless network is a lot like two-way radio communication. Here's what happens:

• A computer's wireless adapter translates data into a radio signal and transmits it using an antenna.

• A wireless router receives the signal and decodes it. The router sends the information to the Internet using a physical, wired Ethernet① connection.

The process also works in reverse, with the router receiving information from the Internet, translating it into a radio signal and sending it to the computer's wireless adapter.

The radios used for WiFi communication are very similar to the radios used for walkie talkies, cell phones and other devices. They can transmit and receive radio waves, and they can convert 1 s and 0 s into radio waves and convert the radio waves back into 1s and 0s. But WiFi radios have a few notable differences from other radios:

• They transmit at frequencies of 2.4 GHz or 5 GHz. This frequency is considerably higher than the frequencies used for cell phones, walkie-talkies and televisions. The higher frequency allows the signal to carry more data.

• They use 802.11 networking standards, which come in several flavors: 802.11a transmits at 5 GHz and can move up to 54 megabits of data per second. It also uses Orthogonal Frequency-Division Multiplexing (OFDM)②, a more efficient coding technique that splits that radio signal into several sub-signals before they reach a receiver. This greatly reduces interference. 802.11b is the slowest and least expensive standard. For a while, its cost made it popular, but now it's becoming less common as faster standards become less expensive. 802.11b transmits in the 2.4 GHz frequency band of the radio spectrum. It can handle up to 11 megabits of data per second, and it uses Complementary Code Keying (CCK)③ modulation to improve speeds. 802.11g transmits at 2.4 GHz like 802.11b, but it's a lot faster—it can handle up to 54 megabits of data per second. 802.11g is faster because it uses the same OFDM coding as 802.11a. 802.11n is the newest standard that is widely available. This standard significantly improves speed and range. For instance, although 802.11g theoretically moves 54 megabits of data per

---

① A Local-Area Network (LAN) architecture developed by Xerox Corporation in cooperation with DEC and Intel in 1976. Ethernet uses a <u>bus or star topology</u> (总线或星型拓扑) and supports data transfer rates of 10 Mbps. The Ethernet specification served as the basis for the IEEE 802.3 standard, which specifies the physical and lower software layers. Ethernet uses the <u>CSMA/CD</u> (带有冲突检测的载波侦听多路存取, Carrier Sense Multiple Access with Collision Detection) access method to handle simultaneous demands. It is one of the most widely implemented LAN standards.

② Frequency-Division Multiplexing (FDM) is a form of signal <u>multiplexing</u> (['mʌltipleksiŋ]n. 多路技术) which involves assigning non-overlapping frequency ranges to different signals or to each "user" of a medium.

③ Complementary Code Keying (CCK) is a modulation scheme used with Wireless Local Area Networks (WLANs) that employ the IEEE 802.11b specification.

second, it only achieves real-world speeds of about 24 megabits of data per second because of network congestion. 802.11n, however, reportedly can achieve speeds as high as 140 megabits per second.

• Other 802.11 standards focus on specific applications of wireless networks, like Wide Area Networks (WANs) inside vehicles or technology that lets you move from one wireless network to another seamlessly.

• WiFi radios can transmit on any of three frequency bands. Or, they can "frequency hop" rapidly between the different bands. Frequency hopping helps reduce interference and lets multiple devices use the same wireless connection simultaneously.

As long as they all have wireless adapters, several devices can use one router to connect to the Internet. This connection is convenient, virtually invisible and fairly reliable; however, if the router fails or if too many people try to use high-bandwidth applications at the same time, users can experience interference or lose their connections.

2. WiFi Hotspots

If you want to take advantage of public WiFi hotspots or start a wireless network in your home, the first thing you'll need to do is make sure your computer has the right gear. Most new laptops and many new desktop computers come with built-in wireless transmitters. If your laptop doesn't, you can buy a wireless adapter that plugs into the PC card slot or USB① port(See Figure 7.2). Desktop computers can use USB adapters, or you can buy an adapter that plugs into the PCI slot inside the computer's case. Many of these adapters can use more than one 802.11 standard.

Wireless adapters can plug into a computer's PC card slot or USB port.

Figure 7.2 Wireless a dapters

---

① Universal Serial Bus (USB) is an <u>industry standard</u> (产业标准，行业标准) developed in the mid-1990s that defines the cables, <u>connectors</u> ([kə'nəktə]n. 连接器) and communications protocols used in a bus for connection, communication and <u>power supply</u> (电源) between computers and electronic devices. USB was designed to standardize the connection of computer peripherals, such as keyboards, <u>pointing devices</u> (定点设备), digital cameras, printers, portable media players, disk drives and network adapters to personal computers, both to communicate and to supply electric power. It has become <u>commonplace</u> (['kɔmənpleis]n. 平凡的事) on other devices, such as smartphones, PDAs and video game consoles. USB has effectively replaced a variety of earlier interfaces, such as serial and parallel ports, as well as separate power chargers for portable devices.

Once you've installed your wireless adapter and the drivers that allow it to operate, your computer should be able to automatically discover existing networks. This means that when you turn your computer on in a WiFi hotspot, the computer will inform you that the network exists and ask whether you want to connect to it. If you have an older computer, you may need to use a software program to detect and connect to a wireless network.

Being able to connect to the Internet in public hotspots is extremely convenient. Wireless home networks are convenient as well. They allow you to easily connect multiple computers and to move them from place to place without disconnecting and reconnecting wires.

3. Building a Wireless Network

If you already have several computers networked in your home, you can create a wireless network with a wireless access point. If you have several computers that are not networked, or if you want to replace your Ethernet network[①], you'll need a wireless router. This is a single unit that contains:

- A port to connect to your cable or DSL modem;
- A router;
- An Ethernet hub;
- A firewall;
- A wireless access point.

A wireless router allows you to use wireless signals or Ethernet cables to connect your computers to one another, to a printer and to the Internet. Most routers provide coverage for about 100 feet (30.5 meters) in all directions, although walls and doors can block the signal. If your home is very large, you can buy inexpensive range extenders or repeaters to increase your router's range.

As with wireless adapters, many routers can use more than one 802.11 standard. 802.11b routers are slightly less expensive, but because the standard is older, they're slower than 802.11a, 802.11g and 802.11n routers. Most people select the 802.11g option for its speed and reliability.

Once you plug in your router, it should start working at its default settings(See Figure 7.3). Most routers let you use a Web interface to change your settings. You can select:

- The name of the network, known as its Service Set IDentifier (SSID)[②]—The default setting is usually the manufacturer's name.
- The channel that the router uses—Most routers use channel 6 by default. If you live in an apartment and your neighbors are also using channel 6, you may experience interference. Switching to a different channel should eliminate the problem.

---

① Ethernet is a family of computer networking technologies for Local Area Networks (LANs) commercially introduced in 1980. Standardized in IEEE 802.3, Ethernet has largely replaced competing wired LAN technologies.

② A Service Set IDentifier (SSID) is a sequence of characters that uniquely names a Wireless Local Area Network (WLAN). An SSID is sometimes referred to as a "network name". This name allows stations to connect to the desired network when multiple independent networks operate in the same physical area.

A wireless router uses an antenna to send signals to wireless devices and a wire to send signals to the Internet.

Figure 7.3　Wireless router

• Your router's security options—Many routers use a standard, publicly available sign-on, so it's a good idea to set your own username and password.

Security is an important part of a home wireless network, as well as public WiFi hotspots. If you set your router to create an open hotspot, anyone who has a wireless card will be able to use your signal. Most people would rather keep strangers out of their network, though. Doing so requires you to take a few security precautions.

It's also important to make sure your security precautions are current. The Wired Equivalency Privacy (WEP) security measure was once the standard for WAN security. The idea behind WEP was to create a wireless security platform that would make any wireless network as secure as a traditional wired network. But hackers discovered vulnerabilities in the WEP approach, and today it's easy to find applications and programs that can compromise a WAN running WEP security.

To keep your network private, you can use one of the following methods:

• WiFi Protected Access (WPA)[①] is a step up from WEP and is now part of the 802.11i wireless network security protocol. It uses Temporal Key Integrity Protocol (TKIP) encryption. As with WEP, WPA security involves signing on with a password. Most public hotspots are either open or use WPA or 128-bit WEP technology, though some still use the vulnerable WEP approach.

• Media Access Control (MAC)[②] address filtering is a little different from WEP or WPA. It

---

① WiFi Protected Access (WPA) and WiFi Protected Access II (WPA2) are two security protocols and security certification programs developed by the WiFi Alliance ([əˈlaiəns]n. 联盟，联合) to secure wireless computer networks. The Alliance defined these in response to serious weaknesses researchers had found in the previous system, WEP.

② The Media Access Control (n.MAC) data communication protocol sub-layer, also known as the medium access control, is a sublayer ([ˈsʌbˈleiə]下层，低层，次层，内层) of the data link layer (数据链路层) specified in the seven-layer OSI model (layer 2). It provides addressing and channel access control mechanisms that make it possible for several terminals or network nodes to communicate within a multiple access network that incorporates a shared medium, e.g. Ethernet. The hardware that implements the MAC is referred to as a medium access controller.

doesn't use a password to authenticate users—it uses a computer's physical hardware. Each computer has its own unique MAC address. MAC address filtering allows only machines with specific MAC addresses to access the network. You must specify which addresses are allowed when you set up your router. This method is very secure, but if you buy a new computer or if visitors to your home want to use your network, you'll need to add the new machines' MAC addresses to the list of approved addresses. The system isn't foolproof. A clever hacker can spoof a MAC address—that is, copy a known MAC address to fool the network that the computer he or she is using belongs on the network.

Wireless networks are easy and inexpensive to set up, and most routers' Web interfaces are virtually self-explanatory.

New Words

widespread	['waidspred]	adj. 分布广泛的,普遍的
unobtrusive	[ˌʌnəb'truːsiv]	adj. 不引人注目的
hotspot	['hɔtspɔt]	n. 热点,热区
adapter	[ə'dæptə]	n. 适配器
decode	[ˌdiː'kəud]	vt. 解码,译解
considerably	[kən'sidərəbəli]	adv. 相当地
frequency	['friːkwənsi]	n. 频率,周率
split	[split]	v. (使)裂开,分裂,分离
sub-signal	[sʌb-'signl]	n. 子信号
interference	[ˌintə'fiərəns]	n. 冲突,干涉
modulation	[ˌmɔdju'leiʃən]	n. 调制
draft	[drɑːft]	n. 草稿,草案,草图
		v. 草拟
ratify	['rætifai]	vt. 批准,认可
band	[bænd]	n. 波段
simultaneously	[siməl'teiniəsli]	adv. 同时地
bandwidth	['bændwidθ]	n. 带宽
gear	[giə]	v. 调整,(使)适合
		n. 装置
detect	[di'tekt]	vt. 探测,发觉
disconnecting	[ˌdiskə'nektiŋ]	n. 拆开,解脱,分离
reconnect	[ˌriːkə'nekt]	v. 再连接
hub	[hʌb]	n. 网络集线器,网络中心
printer	['printə]	n. 打印机
coverage	['kʌvəridʒ]	n. 覆盖

eliminate	[iˈlimineit]	vt. 排除，消除
precaution	[priˈkɔːʃən]	n. 预防，警惕，防范
vulnerability	[ˌvʌlnərəˈbiləti]	n. 弱点，攻击
filter	[ˈfiltə]	n. 滤波器，过滤器，滤光器；筛选
		vt. 过滤，渗透
approved	[əˈpruːvd]	adj. 经核准的，被认可的
foolproof	[ˈfuːlpruːf]	adj. 十分简单的，十分安全的，极坚固的
hacker	[ˈhækə]	n. 电脑黑客
spoof	[spuːf]	v. 哄骗
self-explanatory	[self-ikˈsplænətəri]	adj. 自明的，自解释的

✎ Phrases

in the middle of …	在……的中间
wireless network	无线网络
on the lookout for	寻找，注意
radio waves	无线电波
in reverse	反过来，与……相反
walkie talkie	携带式无线电话机，步话机
split … into …	把……分为……
frequency hop	跳频
in all directions	四面八方，各方面
default setting	默认设置
Service Set IDentifier (SSID)	服务集标识符
by default	默认

✎ Abbreviations

WiFi (Wireless Fidelity)	无线局域网
OFDM (Orthogonal Frequency-Division Multiplexing)	正交频分复用
CCK (Complementary Code Keying)	补码键控
WAN (Wide Area Network)	广域网
PCI (Peripheral Component Interconnect)	外设部件互连标准
DSL (Digital Subscriber Line)	数字用户线路
WEP (Wired Equivalency Privacy)	有线对等隐私，有线对等加密
WPA (WiFi Protected Access)	无线安全访问
TKIP (Temporal Key Integrity Protocol)	暂时密钥集成协议，临时密钥完整协议

Exercises

I. Answer the following questions according to the text.

1. What are the radios used for WiFi communication very similar to?
2. What frequencies do WiFi radios transmit at? What standards do they use networking?
3. What happens if the router fails or if too many people try to use high-bandwidth applications at the same time?
4. What can you do if your laptop doesn't have built-in wireless transmitters?
5. When can you create a wireless network with a wireless access point?
6. What does a wireless router contain?
7. What does a wireless router allow you to do?
8. What happens if you set your router to create an open hotspot?
9. What is WiFi Protected Access? What does it do to keep your network private?
10. What can a clever hacker do?

II. Translate the following terms or phrases from English into Chinese and vice versa.

1. frequency hop
2. wireless network
3. n. 频率，周率
4. detect
5. 默认设置
6. n. 带宽
7. decode
8. adapter
9. interference
10. hub

III. Fill in the blanks with the words given below.

| range | trademark | smartphone | attacker | products |
| updated | encryption | exchange | block | synonym |

◇◇◇ **WiFi** ◇◇◇

WiFi is a popular technology that allows an electronic device to ___1___ data wirelessly (using radio waves) over a computer network, including high-speed Internet connections. The WiFi Alliance defines WiFi as any "wireless local area network (WLAN) ___2___ that are based on the Institute of Electrical and Electronics Engineers' (IEEE) 802.11 standards". However, since most modern WLANs are based on these standards, the term "WiFi" is used in general English as a ___3___ for "WLAN".

A device that can use WiFi (such as a personal computer, video game console, 4 , tablet, or digital audio player) can connect to a network resource such as the Internet via a wireless network access point. Such an access point (or hotspot) has a 5 of about 20 meters (65 feet) indoors and a greater range outdoors. Hotspot coverage can comprise an area as small as a single room with walls that 6 radio waves or as large as many square miles — this is achieved by using multiple overlapping access points.

"WiFi" is a 7 of the WiFi Alliance and the brand name for products using the IEEE 802.11 family of standards. Only WiFi products that complete WiFi Alliance interoperability certification testing successfully may use the "WiFi CERTIFIED" designation and trademark.

WiFi has had a checkered security history. Its earliest 8 system, WEP, proved easy to break. Much higher quality protocols, WPA and WPA2, were added later. However, an optional feature added in 2007, called WiFi Protected Setup (WPS), has a flaw that allows a remote 9 to recover the router's WPA or WPA2 password in a few hours on most implementations. Some manufacturers have recommended turning off the WPS feature. The WiFi Alliance has since 10 its test plan and certification program to ensure all newly-certified devices resist brute-force AP PIN attacks.

IV. Translate the following passage from English to Chinese.

✧✧✧ Advantages of WiFi ✧✧✧

WiFi allows cheaper deployment of Local Area Networks (LANs). Also spaces where cables cannot be run, such as outdoor areas and historical buildings, can host wireless LANs.

Manufacturers are building wireless network adapters into most laptops. The price of chipsets for WiFi continues to drop, making it an economical networking option included in even more devices.

Different competitive brands of access points and client network-interfaces can inter-operate at a basic level of service. Products designated as "WiFi Certified" by the WiFi Alliance are backwards compatible. Unlike mobile phones, any standard WiFi device will work anywhere in the world.

The current version of WiFi Protected Access encryption (WPA2) is widely considered secure, provided users employ a strong passphrase. New protocols for quality-of-service (WMM) make WiFi more suitable for latency-sensitive applications (such as voice and video); and power saving mechanisms (WMM Power Save) improve battery operation.

Bluetooth

Bluetooth is a proprietary open wireless technology standard for exchanging data over short distances (using short wavelength radio transmissions in the ISM band from 2400–2480 MHz) from fixed and mobile devices, creating Personal Area Networks① (PANs) with high levels of security. Created by telecoms vendor Ericsson in 1994, it was originally conceived as a wireless alternative to RS-232② data cables. It can connect several devices, overcoming problems of synchronization (See figure 7.4).

Figure 7.4　Bluetooth Logo

1. Implementation

Bluetooth uses a radio technology called frequency-hopping spread spectrum③, which chops up the data being sent and transmits chunks of it on up to 79 bands (1 MHz each; centered from 2402 to 2480 MHz) in the range 2,400–2,483.5 MHz (allowing for guard bands). This range is in the globally unlicensed Industrial, Scientific and Medical (ISM)④ 2.4 GHz short-range radio frequency band.

Originally Gaussian Frequency-Shift Keying (GFSK) modulation was the only modulation scheme available; subsequently, since the introduction of Bluetooth 2.0+EDR, π/4-DQPSK and

---

① A Personal Area Network (PAN) is a computer network used for communication among computerized devices, including telephones and personal digital assistants. PANs can be used for communication among the personal devices themselves (intrapersonal communication), or for connecting to a higher level network and the Internet (an uplink). A Wireless Personal Area Network (WPAN) is a PAN carried over wireless network technologies such as IrDA (Infrared Data Association,红外数据协会), Bluetooth, Wireless USB, Z-Wave, ZigBee, or even Body Area Network. The reach of a WPAN varies from a few centimeters to a few meters. A PAN may also be carried over wired computer buses such as USB and FireWire.

② In telecommunications, RS-232 (Recommended Standard 232) is the traditional name for a series of standards for serial binary single-ended data and control signals connecting between a DTE (Data Terminal Equipment，数据终端设备) and a DCE (Data Circuit-terminating Equipment，数据电路终端设备). It is commonly used in computer serial ports. The standard defines the electrical characteristics and timing of signals, the meaning of signals, and the physical size and pin out of connectors.

③ Frequency-Hopping Spread Spectrum (FHSS) is a method of transmitting radio signals by rapidly switching a carrier among many frequency channels, using a pseudorandom ([ˌpsjuːdəʊˈrændəm]adj. 伪随机的) sequence known to both transmitter and receiver. It is utilized as a multiple access method in the Frequency-Hopping Code Division Multiple Access (FH-CDMA，跳频码分多址) scheme.

④ The Industrial, Scientific and Medical (ISM) radio bands are radio bands (portions of the radio spectrum) reserved internationally for the use of Radio Frequency (RF) energy for industrial, scientific and medical purposes other than communications.

8DPSK modulation may also be used between compatible devices. Devices functioning with GFSK are said to be operating in Basic Rate (BR) mode where an instantaneous data rate of 1 Mbit/s is possible. The term Enhanced Data Rate (EDR) is used to describe π/4-DPSK and 8DPSK schemes, each giving 2 and 3 Mbit/s respectively. The combination of these (BR and EDR) modes in Bluetooth radio technology is classified as a "BR/EDR radio".

Bluetooth is a packet-based protocol with a master-slave structure. One master may communicate with up to 7 slaves in a piconet①; all devices share the master's clock. Packet exchange is based on the basic clock, defined by the master, which ticks at 312.5 μs intervals. Two clock ticks make up a slot of 625 μs; two slots make up a slot pair of 1250 μs. In the simple case of single-slot packets the master transmits in even slots and receives in odd slots; the slave, conversely, receives in even slots and transmits in odd slots. Packets may be 1, 3 or 5 slots long but in all cases the master transmit will begin in even slots and the slave transmit in odd slots.

Bluetooth provides a secure way to connect and exchange information between devices such as faxes, mobile phones, telephones, laptops, personal computers, printers, Global Positioning System (GPS)② receivers, digital cameras, and video game consoles.

A master Bluetooth device can communicate with a maximum of seven devices in a piconet (an ad-hoc computer network using Bluetooth technology), though not all devices support this limit. The devices can switch roles, by agreement, and the slave can become the master (for example, a headset initiating a connection to a phone will necessarily begin as master, as initiator of the connection; but may subsequently prefer to be slave).

The Bluetooth Core Specification provides for the connection of two or more piconets to form a scatternet③, in which certain devices simultaneously play the master role in one piconet and the slave role in another.

At any given time, data can be transferred between the master and one other device (except for the little-used broadcast mode). The master chooses which slave device to address; typically, it switches rapidly from one device to another in a round-robin fashion. Since it is the master that chooses which slave to address, whereas a slave is (in theory) supposed to listen in each receive slot, being a master is a lighter burden than being a slave. Being a master of seven slaves is possible; being a slave of more than one master is difficult. The specification is vague as to

---

① The original piconet was a networking type used on RM Nimbus computers. These days, a "piconet" is an ad-hoc computer network linking a user group of devices using Bluetooth technology protocols to allow one master device to interconnect with up to seven active slave devices (because a three-bit MAC address is used). Up to 255 further slave devices can be inactive, or parked, which the master device can bring into <u>active status</u> (活动状态) at any time.
Piconet range will vary according to the class of the Bluetooth device. Data transfer rates vary between about 200 and 2100 kbit/s at the application.

② The Global Positioning System (GPS) is a space-based <u>satellite</u> (['sætəlait]n. 人造卫星) navigation system that provides location and time information in all weather, anywhere on or near the Earth, where there is an unobstructed line of sight to four or more GPS satellites.

③ A scatternet is a type of ad-hoc computer network consisting of two or more piconets.

required behaviour in scatternets.

2. Uses

Bluetooth is a standard wire-replacement communications protocol primarily designed for low power consumption, with a short range (power-class-dependent, but effective ranges vary in practice; see table below) based on low-cost transceiver microchips in each device. Because the devices use a radio (broadcast) communications system, they do not have to be in visual line of sight of each other, however a quasi optical wireless path must be viable.

Class	Maximum permitted power		Range/m
	mW	dBm	
Class 1	100	20	~100
Class 2	2.5	4	~10
Class 3	1	0	~5

The effective range varies due to propagation conditions, material coverage, production sample variations, antenna configurations and battery conditions. In most cases the effective range of class 2 devices is extended if they connect to a class 1 transceiver, compared to a pure class 2 network. This is accomplished by the higher sensitivity and transmission power of Class 1 devices.

Version	Data rate	Maximum application throughput
Version 1.2	1 Mb/s	0.7 Mb/s
Version 2.0 + EDR	3 Mb/s	2.1 Mb/s
Version 3.0 + HS	Up to 24 Mb/s	See Version 3.0+HS.
Version 4.0	Up to 24 Mb/s	See Version 3.0+HS.

While the Bluetooth Core Specification does mandate minimums for range, the range of the technology is application specific and is not limited. Manufacturers may tune their implementations to the range needed to support individual use cases.

2.1 Bluetooth profiles

To use Bluetooth wireless technology, a device has to be able to interpret certain Bluetooth profiles, which are definitions of possible applications and specify general behaviors that Bluetooth enabled devices use to communicate with other Bluetooth devices. These profiles include settings to parameterize and to control the communication from start. Adherence to profiles saves the time for transmitting the parameters anew before the bi-directional link becomes effective. There are a wide range of Bluetooth profiles that describe many different types of applications or use cases for devices.

2.2 List of applications

A typical Bluetooth mobile phoneheadset.

• Wireless control of and communication between a mobile phone and a handsfree headset. This was one of the earliest applications to become popular.

• Wireless control of and communication between a mobile phone and a Bluetooth compatible car stereo system.

• Wireless Bluetooth headset and Intercom.

• Wireless networking between PCs in a confined space and where little bandwidth is required.

• Wireless communication with PC input and output devices, the most common being the mouse, keyboard and printer.

• Transfer of files, contact details, calendar appointments, and reminders between devices with OBEX[①].

• Replacement of previous wired RS-232 serial communications in test equipment, GPS receivers, medical equipment, bar code scanners, and traffic control devices.

• For controls where infrared was often used.

• For low bandwidth applications where higher USB bandwidth is not required and cable-free connection desired.

• Sending small advertisements from Bluetooth-enabled advertising hoardings to other, discoverable, Bluetooth devices.

• Wireless bridge between two Industrial Ethernet networks.

• Three seventh-generation game consoles, Nintendo's Wii and Sony's PlayStation 3 and PSP Go, use Bluetooth for their respective wireless controllers.

• Dial-up internet access on personal computers or PDAs using a data-capable mobile phone as a wireless modem.

• Short range transmission of health sensor data from medical devices to mobile phone, set-top box[②] or dedicated telehealth devices.

• Allowing a DECT[③] phone to ring and answer calls on behalf of a nearby cell phone.

• Real-Time Location Systems (RTLS), are used to track and identify the location of objects in real-time using "Nodes" or "tags" attached to, or embedded in the objects tracked, and "Readers" that receive and process the wireless signals from these tags to determine their

---

① OBEX (also termed IrOBEX) is a communications protocol that facilitates the exchange of binary objects between devices.
② A Set-Top Box (STB) or Set-Top Unit (STU) is an information appliance device that generally contains a tuner and connects to a television set and an external source of signal, turning the signal into content which is then displayed on the television screen or other display device. Set-top boxes are used in cable television and satellite television systems, to transform the signal from the cable or satellite to a form that can be used by the television set or other receiver.
③ Digital Enhanced Cordless Telecommunications, usually known by the acronym DECT, is a digital communication standard, which is primarily used for creating cordless phone systems.

locations.

　　• Personal security application on mobile phones for prevention of theft or loss of items. The protected item has a Bluetooth marker (e.g. a tag) that is in constant communication with the phone. If the connection is broken (the marker is out of range of the phone) then an alarm is raised.

　　2.3　Bluetooth vs. WiFi (IEEE 802.11)

　　Bluetooth and WiFi (the brand name for products using IEEE 802.11 standards) have some similar applications: setting up networks, printing, or transferring files. WiFi is intended as a replacement for cabling for general local area network access in work areas. This category of applications is sometimes called Wireless Local Area Networks (WLAN). Bluetooth was intended for portable equipment and its applications. The category of applications is outlined as the Wireless Personal Area Network (WPAN). Bluetooth is a replacement for cabling in a variety of personally carried applications in any setting and can also support fixed location applications such as smart energy functionality in the home (thermostats, etc.).

　　WiFi is a wireless version of a common wired Ethernet network, and requires configuration to set up shared resources, transmit files, and to set up audio links (for example, headsets and hands-free devices). WiFi uses the same radio frequencies as Bluetooth, but with higher power, resulting in higher bit rates and better range from the base station. The nearest equivalents in Bluetooth are the DUN[①] profile, which allows devices to act as modem interfaces, and the PAN profile, which allows for ad-hoc networking.

　　2.4　Devices

　　A Bluetooth USB dongle with a 100 m range. The MacBook Pro, shown, also has a built in Bluetooth adaptor.

　　Bluetooth exists in many products, such as the iPod Touch, Lego Mindstorms NXT, PlayStation 3, PSP Go, telephones, the Nintendo Wii, and some high definition headsets, modems, and watches. The technology is useful when transferring information between two or more devices that are near each other in low-bandwidth situations. Bluetooth is commonly used to transfer sound data with telephones (i.e., with a Bluetooth headset) or byte data with hand-held computers (transferring files).

　　Bluetooth protocols simplify the discovery and setup of services between devices.

New Words

Bluetooth　　　　['bluːtuːθ]　　　　　　n. 蓝牙

[①] This profile provides a standard to access the Internet and other dial-up services over Bluetooth. The most common scenario is accessing the Internet from a laptop by dialing up on a mobile phone, wirelessly. It is based on Serial Port Profile (SPP, 串行协议), and provides for relatively easy conversion of existing products, through the many features that it has in common with the existing wired serial protocols for the same task.

wavelength	[ˈweivleŋθ]	n. 波长
telecom	[ˈteləkɔm]	n. 电信
instantaneous	[ˌinstənˈteinjəs]	adj. 瞬间的，即刻的，即时的
describe	[disˈkraib]	v. 描述
respectively	[riˈspektivli]	adv. 分别地，各个地
piconet	[ˈpikənet]	n. 微微网
interval	[ˈintəvəl]	n. 时间间隔
console	[kənˈsəul]	n. 控制台
headset	[ˈhedset]	n. 戴在头上的耳机或听筒；头套
scatternet	[ˈskætənet]	n. 分布网，分散网
broadcast	[ˈbrɔːdkɑːst]	n. & v. 广播
mode	[məud]	n. 模式，方式，样式
burden	[ˈbəːdn]	n. 负担
vague	[veig]	adj. 含糊的，不清楚的
transceiver	[trænˈsiːvə]	n. 收发器
quasi	[ˈkwɑːziː(ː), ˈkweisai]	adj. 类似的，准的
sensitivity	[ˌsensiˈtiviti]	n. 灵敏度，灵敏性
mandate	[ˈmændeit]	n. 命令，要求
parameterize	[pəˈræmitəraiz]	vt. 确定……的参数，用参数表示
adherence	[ədˈhiərəns]	n. 依附，黏着
parameter	[pəˈræmitə]	n. 参数，参量
intercom	[ˈintəkɔm]	n. 联络所用对讲电话装置，内部通信联络系统
confined	[kənˈfaind]	adj. 被限制的，狭窄的
infrared	[ˈinfrəˈred]	adj. 红外线的 n. 红外线
Nintendo	[ninˈtendəu]	n. 任天堂(游戏名)
thermostat	[ˈθəːməstæt]	n. 自动调温器，温度调节装置
definition	[ˌdefiˈniʃən]	n. 精确度，清晰度
byte	[bait]	n. 字节

✎ Phrases

short distance	短距离
data cable	数据传输电缆
Frequency-Hopping Spread Spectrum (FHSS)	跳频扩频
chop up	切开，切细
Gaussian Frequency-Shift Keying (GFSK)	高斯频移键控

Basic Rate (BR)	基本速率
packet-based protocol	基于分组的协议
master-slave structure	主-从结构
master clock	主时钟
clock tick	时钟节拍，时钟周期
Global Positioning System (GPS)	全球定位系统
round-robin fashion	轮流的方式
power consumption	能量消耗，功率消耗
propagation condition	传播条件
bi-directional link	双向链接
become effective	生效
stereo system	立体音响系统
serial communication	串行通信
Eethernet network	以太网
Set-Top Box (STB)	机顶盒
Real-Time Location Systems (RTLS)	实时定位系统
portable equipment	可携带设备
hand-held computer	手持式计算机

Abbreviations

ISM (Industrial, Scientific and Medical)	工业，科学和医学
DQPSK (Differential Quadrature Reference Phase Shift Keying)	差分四相相移键控
EDR (Enhanced Data Rate)	增强数据率
8DPSK (Differential encoded 8-ary Phase Shift Keying)	差分编码八元相移键控
dBm	毫瓦分贝
OBEX (OBject EXchange)	对象交换
DECT (Digital Enhanced Cordless Telecommunications)	数字增强无绳电话
DUN (Dial-Up Networking)	拨号网络应用

Exercises

I. Answer the following questions according to the text.

　　1. What is Bluetooth?

　　2. What is the radio technology Bluetooth uses called?

3. How many slaves may one master communicate with in a piconet? What do all devices share?

4. In the simple case of single-slot packets how does the master transmit and receives?

5. Why don't the devices have to be in visual line of sight of each other?

6. What are Bluetooth profiles?

7. What do these profiles include?

8. What is WiFi intended as?

9. What is Bluetooth intended for?

10. Please list the products Bluetooth exists.

II. Translate the following terms or phrases from English into Chinese and vice versa.

1. parameterize
2. transceiver
3. master clock
4. scatternet
5. 能量消耗，功率消耗
6. piconet
7. synchronization
8. sensitivity
9. n. 波长
10. serial communication

Text A 参考译文

WiFi 是如何工作的？

如果你已经去过机场、咖啡厅、图书馆或宾馆，你很可能已经身在无线网络之中。许多人在家里使用无线网络(被称为"WiFi 或 802.11 联网")来连接计算机。一些城市也使用该技术为市民提供免费或低价的因特网访问。在不久的将来，无线网络的应用将会非常广泛，你可以不用网线而随时随地访问因特网(如图 7.1 所示，图略)。

WiFi 有许多优点：无线网络容易建立，资费也便宜；它们不引人注意——除非你用笔记本电脑刻意寻找，否则当你在热区使用时也感觉不到它的存在。在本文中，我们将研究这种让信息在空中传输的技术，也会讨论在你家中建立无线网络所需的条件。

1. 什么是 WiFi？

无线网络使用无线电波传播，就如同手机、电视和收音机一样。实际上，通过无线网络的通信与双向无线电通信很像，过程如下：
- 计算机的无线适配器把数据转换为无线电信号并用天线发射出去。
- 无线路由器接收到这个信号并解码，然后路由器把信息发送给使用物理的、有线的以太网连接的因特网。

这个过程也可以反过来，路由器从因特网接收信息，然后转换为无线电信号并发送给计算机的无线适配器。

用于 WiFi 通信的无线电与用于步话机、手机和其他设备的无线电几乎一样，它们可以发射和接收无线电波，并可以把一系列的 1 和 0 转换为无线电波，也能把无线电波反过来转换为一系列的 1 和 0。但 WiFi 无线电通信与其他无线电也有一些明显的不同：
- 它们以 2.4 GHz 或 5 GHz 的频率传输。这个频率比手机、步话机和电视所用的频率高很多。这种较高的频率允许信号携带更多的数据。
- 它们使用 802.11 联网标准，802.11 联网标准有以下几个部分：802.11a 以 5 GHz 的频率传输并且每秒可以传输高达 54 兆位的数据；它也使用正交频分复用——一种更高效的编码技术——该技术在无线电信号到达接收器之前把它们分为几个子信号，这极大地减少了冲突。802.11b 是最慢和最便宜的标准，一度因为便宜而流行。但是，因为更快的标准现在也变得便宜，它就不那么常用了。802.11b 以 2.4 GHz 频率的无线电射频波段传输，每秒可以处理 11 兆位信息并使用补码键控调制技术来增加速度。802.11g 与 802.11b 一样以 2.4 GHz 传输，但快得多——它每秒可以处理 54 兆位的数据。802.11g 更快的原因是它与 802.11a 一样使用 OFDM 译码技术。802.11n 是广泛使用的最新标准。该标准明显改进了速度和范围。例如，虽然 802.11g 理论上每秒可以移动 54 兆位的数据，但由于网络阻塞，所以实际速度只有每秒 24 兆位。但据估计，802.11n 可以达到每秒 140 兆位的高速。
- 其他 802.11 标准注重无线网络的特殊应用。像车辆中的广域网一样，该技术可以实现从一个无线网络到另一个无线网络的无缝切换。
- WiFi 无线电可以使用三种频率中的任何一个波段来发射，或者在不同波段之间快速"跳频"。跳频可以减少冲突并让多个设备同时使用一个无线连接。

只要都有无线适配器，几个设备就可以使用一个路由器连接到因特网。这些连接不仅方便，而且实际上看不到并相当可靠。但是，如果路由器发生故障或者有太多的人试图同时使用高带宽应用，用户可能会感觉到冲突或丢失连接。

2. WiFi 热区

如果你要使用 WiFi 热区或者在家里启动一个无线网络，首先要确保你的计算机有合适的装置。大多数笔记本计算机和许多台式计算机都带内置的无线传输器。如果你的笔记本计算机没有，你可以购买一个无线适配器并插到 PC 卡插槽中或者 USB 接口上(如图 7.2 所示，图略)。台式计算机则可以使用 USB 适配器，或者买一个适配器插到机箱内的 PCI 槽中。许多适配器可以使用多种 802.11 标准。

一旦你安装了无线适配器并驱动了它,你的计算机就应该能够自动发现现有的网络。这就意味着当在 WiFi 热区中打开计算机时,计算机会告知你网络已经存在并询问你是否连入网络。如果使用的是老式计算机,可能需要使用软件程序来探测并接入无线网络。

在公共热区可以接入因特网是非常方便的,无线家庭网络具有同样的便利性。它们使你可以很容易地接入多台计算机,并且从一个地方转到另一个地方时不用断网再接入。

<u>3. 建立一个无线网络</u>

如果你家里已经有了几个联网的计算机,则你可以建立一个带无线接入点的无线网络。如果家里有几个没有联网的计算机,或者想替换你的以太网,那么你需要一个无线路由器。总的来说,要包括以下各部分:
- 一个连接到电缆或 DSL 调制解调器的端口;
- 一个路由器;
- 一个以太网集线器;
- 一个防火墙;
- 一个无线接入点。

通过无线路由器你可以用无线信号或者电缆把你的计算机与其他计算机、打印机和因特网相连接。尽管墙壁和门会阻挡信号,但大多数路由器可以覆盖方圆 100 英尺(30.5 m)的范围。如果你家比较大,那么可以购买便宜的扩充器或中继器来增大路由器的范围。

就无线适配器而言,许多路由器可以使用多个 802.11 标准。802.11b 路由器稍微便宜点,但因为该标准比较旧,所以速度比 802.11a、802.11g 和 802.11n 路由器慢。许多人因速度和可靠性而选择 802.11g 路由器。

一旦接通路由器,它就按照默认的设置开始工作(如图 7.3 所示,图略)。对大多数路由器而言,你可以通过一个网络界面来改变路由器的设置。你可以进行如下设置:
- 网络名称,被称为服务集标识符(SSID)——默认设置通常为厂家的名字。
- 路由器所用的通道——大多数路由器默认使用通道 6。如果你住在一个单元房中并且你的邻居也使用通道 6,你也许会感觉到冲突。换个通道就可以解决这个问题。
- 路由器的安全选项——许多路由器使用标准的、公共的登录,因此设置自己的名字和密码是个好主意。

安全是家用无线网络的重要部分,公共 WiFi 热区也一样。如果你将自己的路由器设置成一个开放的热区,那么只要有无线网卡的人都可以使用你的信号。许多人更愿意把陌生人拒之门外,因此,你确实需要做一些安全防范。

确保你的安全措施最新也很重要。有线对等加密安全度量曾经是广域网安全标准。有线对等加密观点是建立一个无线安全平台,从而使该平台上的任何无线网络与传统的有线网络一样安全。但黑客们已经发现了有线对等加密安全方法的弱点。如今,很容易找到可以弥补运行有线对等加密的广域网的不足的应用程序。

要保证网络安全,可以采用如下方法:
- 使用 WiFi 网络保护访问来替代 WEP。目前这已经是 802.11i 无线网络安全协议的一部分。它使用了暂时密钥集成协议。就 WEP 而言,WPA 安全使用了带密码的登录。大多数公共热区要么是开放的，要么使用的是 WPA 或者使用的是 128 位的 WEP 技术，但一些人仍然使用易受攻击的 WEP。

• 介质访问控制地址过滤方式与 WEP 或 WPA 有些不同。它不使用密码来鉴别用户——它使用计算机的物理硬件。每个计算机都有自己唯一的 MAC 地址。MAC 地址过滤只允许带有特定 MAC 地址的计算机访问网络。当设置路由器时，必须指定允许访问的地址。这种方法非常安全，但如果你要买一台新计算机或者到你家的访客要使用你的网络，就需要把新计算机的 MAC 地址加到许可地址的列表中。但这个系统也不是十分安全，因为聪明的黑客可以使用假冒的 MAC 地址——即复制一个 MAC 地址来欺骗网络，让网络认为他或她的计算机属于该网络。

无线网络的建立很容易，而且便宜，并且路由器的网页界面实际上是自解释的。

Unit 8 Wireless Sensor Network and It's Application

Text A

Wireless Sensor Network

A Wireless Sensor Network (WSN) consists of spatially distributed autonomous sensors to monitor physical or environmental conditions, such as temperature, sound, vibration, pressure, motion or pollutants and to cooperatively pass their data through the network to a main location. The more modern networks are bidirectional, also enabling control of sensor activity. The development of wireless sensor networks was motivated by military applications such as battlefield surveillance; today such networks are used in many industrial and consumer applications, such as industrial process monitoring and control, machine health monitoring, and so on.

The WSN is built of "nodes"—from a few to several hundreds or even thousands, where each node is connected to one (or sometimes several) sensors. Each such sensor network node has typically several parts: a radio transceiver[①] with an internal antenna or connection to an external antenna, a microcontroller[②], an electronic circuit for interfacing with the sensors and an energy source, usually a battery or an embedded form of energy harvesting. A sensor node[③] might vary in size from that of a shoebox down to the size of a grain of dust. The cost of sensor

---

① A transceiver is a device comprising both a <u>transmitter</u> ([trænz'mɪtə]n. 发射器) and a <u>receiver</u> ([rɪ'siːvə]n. 接受者, 接收器), which are combined and share common circuitry or a single housing. When no circuitry is common between transmit and receive functions, the device is a transmitter-receiver. Technically, transceivers must combine a significant amount of the transmitter and receiver handling circuitry.

② A microcontroller (sometimes abbreviated µC, uC or MCU) is a small computer on a single integrated circuit containing a processor core, memory, and programmable input/output peripherals.

③ A sensor node, also known as a mote, is a node in a wireless sensor network that is capable of performing some processing, gathering <u>sensory</u> (['sensəri]adj. 感觉的) information and communicating with other connected nodes in the network. A mote is a node but a node cannot always be a mote.

nodes is similarly variable, ranging from a few to hundreds of dollars, depending on the complexity of the individual sensor nodes. Size and cost constraints on sensor nodes result in corresponding constraints on resources such as energy, memory, computational speed and communications bandwidth. The topology of the WSNs can vary from a simple star network to an advanced multihop wireless mesh network[①]. The propagation technique between the hops of the network can be routing or flooding.

In computer science and telecommunications, wireless sensor networks are an active research area with numerous workshops and conferences arranged each year.

1. Characteristics

The main characteristics of a WSN include(See Figure 8.1):
- Power consumption constrains for nodes using batteries or energy harvesting;
- Ability to cope with node failures;
- Mobility of nodes;
- Dynamic network topology;
- Communication failures;
- Heterogeneity of nodes;
- Scalability to large scale of deployment;
- Ability to withstand harsh environmental conditions;
- Ease of use;
- Unattended operation;
- Power consumption.

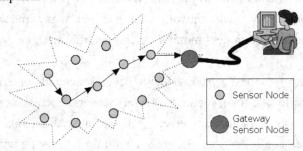

Figure 8.1 Typical multi-hop wireless sensor network architecture

Sensor nodes can be imagined as small computers, extremely basic in terms of their interfaces and their components. They usually consist of a processing unit with limited computational power and limited memory, sensors or MEMS[②] (including specific conditioning circuitry), a communication device (usually radio transceivers or alternatively optical), and a

---

① A Wireless Mesh Network (WMN) is a communications network made up of radio nodes organized in a mesh topology. Wireless Mesh Networks often consist of mesh clients, mesh routers and gateways.
② MEMS (also written as Micro-Electro-Mechanical, MicroElectroMechanical or Microelectronic and MicroElectromechanical Systems) is the technology of very small mechanical devices driven by electricity; it merges at the nano-scale into NanoElectroMechanical Systems (NEMS) and nanotechnology (纳米技术).

power source usually in the form of a battery. Other possible inclusions are energy harvesting modules, secondary ASICs[①], and possibly secondary communication devices (e.g. RS-232 or USB).

The base stations are one or more components of the WSN with much more computational, energy and communication resources. They act as a gateway between sensor nodes and the end user as they typically forward data from the WSN on to a server. Other special components in routing based networks are routers, designed to compute, calculate and distribute the routing tables. Many techniques are used to connect to the outside world including mobile phone networks, satellite phones, radio modems, long-range WiFi links etc. Many base stations are ARM[②]-based running a form of Embedded Linux.

2. Platforms

2.1 Standards and Specifications

Several standards are currently either ratified or under development for wireless sensor networks. There are a number of standardization bodies in the field of WSNs. The IEEE focuses on the physical and MAC[③] layers; the Internet Engineering Task Force(IETF) works on layers 3 and above. In addition to these, bodies such as the International Society of Automation provide vertical solutions, covering all protocol layer. Finally, there are also several non-standard, proprietary mechanisms and specifications.

Standards are used far less in WSNs than in other computing systems which makes most systems incapable of direct communication between different systems. However predominant standards commonly used in WSN communications include:

- WirelessHART[④]
- ISA100
- IEEE 1451
- ZigBee/802.15.4

2.2 Hardware

One major challenge in a WSN is to produce low cost and tiny sensor nodes. There are an increasing number of small companies producing WSN hardware and the commercial situation can be compared to home computing in the 1970s. Many of the nodes are still in the research and

---

① An Application-Specific Integrated Circuit (ASIC) is an Integrated Circuit (IC，集成电路) customized for a particular use, rather than intended for general-purpose use.

② ARM is a 32-bit Reduced Instruction Set Computer (RISC) Instruction Set Architecture (ISA) developed by ARM Holdings. It was named the Advanced RISC Machine and, before that, the Acorn RISC Machine. The ARM architecture is the most widely used 32-bit instruction set (指令集，指令组) architecture in numbers produced.

③ The Media Access Control (MAC) data communication protocol sub-layer, also known as the medium access control, is a sublayer of the data link layer specified in the seven-layer OSI model (layer 2).

④ WirelessHART is a wireless sensor networking technology based on the Highway Addressable Remote Transducer Protocol (HART).

development stage, particularly their software. Also inherent to sensor network adoption is the use very low power methods for data acquisition.

2.3 Software

Energy is the scarcest resource of WSN nodes, and it determines the lifetime of WSNs. WSNs are meant to be deployed in large numbers in various environments, including remote and hostile regions, where ad hoc communications are a key component. For this reason, algorithms and protocols need to address the following issues:

• Lifetime maximization;
• Robustness and fault tolerance;
• Self-configuration.

Some of the important topics in WSN software research are:

• Operating systems;
• Security;
• Mobility;
• Usability—human interface for deployment and management, debugging and end-user control;
• Middleware—the design of middle-level primitives between high level software and the systems.

2.4 Operating systems

Operating systems for wireless sensor network are typically less complex than general-purpose operating systems. They more strongly resemble embedded systems, for two reasons. First, wireless sensor networks are typically deployed with a particular application in mind, rather than as a general platform. Second, a need for low costs and low power leads most wireless sensor nodes to have low-power microcontrollers ensuring that mechanisms such as virtual memory are either unnecessary or too expensive to implement.

It is therefore possible to use embedded operating systems such as eCos or uC/OS for sensor networks. However, such operating systems are often designed with real-time properties.

TinyOS is perhaps the first operating system specifically designed for wireless sensor networks. TinyOS is based on an event-driven programming[①] model instead of multithreading[②]. TinyOS programs are composed of event handler sand tasks with run-to-completion semantics.

---

① In computer programming, event-driven programming or event-based programming is a programming paradigm in which the flow of the program is determined by events—i.e., sensor outputs or user actions (mouse clicks, key presses) or messages from other programs or threads.

② In computer science, a thread of execution is the smallest unit of processing that can be scheduled by an operating system. The implementation of threads and processes differs from one operating system to another, but in most cases, a thread is contained inside a process. Multiple threads can exist within the same process and share resources such as memory, while different processes do not share these resources. In particular, the threads of a process share the latter's <u>instructions</u> ([inˈstrʌkʃən]n. 指令) (its code) and its context (the values that its variables reference <u>at any given moment</u> (随时)).

When an external event occurs, such as an incoming data packet or a sensor reading, TinyOS signals the appropriate event handler to handle the event. Event handlers can post tasks that are scheduled by the TinyOS kernel some time later.

LiteOS is a newly developed OS for wireless sensor networks, which provides UNIX-like abstraction and support for the C programming language.

3. Simulation of WSNs

In general, there are two ways to develop simulations of WSNs. Either use a custom platform to develop the simulation. And the second option is to develop one's own simulation.

3.1 Simulators

As such, at present Agent-based Modeling and Simulation is the only paradigm which allows the simulation of even complex behavior in the environments of Wireless sensors (such as flocking).

Network Simulators like QualNet, NetSim, and NS2 can be used to simulate Wireless Sensor Network.

3.2 Agent-based simulation of WSN

Agent-based simulation of wireless sensor and ad hoc networks is a relatively newer paradigm. Agent-based modelling was originally based on social simulation. A recent article on agent-based simulation published in the IEEE Communications magazine gives examples and tutorials on how to develop custom agent-based simulation models for wireless sensors, mobile robots and P2P networks in a short period of time (few hours). A formal agent-based simulation framework using formal specification using Z notation demonstrating the use of agent-based modeling to represent simulation of complex behavior in the environment of sensors is given in. Agent-based simulation has also been shown to be useful for modeling and simulation for quantifying emergent behavior in the vicinity of WSN nodes.

4. Other concepts

4.1 Distributed sensor network

If a centralised architecture is used in a sensor network and the central node fails, then the entire network will collapse, however the reliability of the sensor network can be increased by using a distributed control architecture.

Distributed control is used in WSNs for the following reasons:
- Sensor nodes are prone to failure;
- For better collection of data;
- To provide nodes with backup in case of failure of the central node.

There is also no centralised body to allocate the resources and they have to be self organised.

4.2 Data inte

gration and sensor web

The data gathered from wireless sensor networks is usually saved in the form of numerical data in a central base station. Additionally, the Open Geospatial Consortium (OGC)[①] is specifying standards for interoperability interfaces and metadata encodings that enable real time integration of heterogeneous sensor webs into the Internet, allowing any individual to monitor or control Wireless Sensor Networks through a Web Browser.

4.3 In-network processing

To reduce communication costs some algorithms remove or reduce nodes redundant sensor information and avoid forwarding data that is of no use. As nodes can inspect the data they forward they can measure averages or directionality, for example, of readings from other nodes.

4.4 Information fusion

In wireless sensor networks, information fusion, also called data fusion, has been developed for processing sensor data by filtering, aggregating, and making inferences about the gathered data. Information fusion deals with the combination of multiple sources to obtain improved information: cheaper, greater quality or greater relevance. Within the wireless sensor networks domain, simple aggregation techniques such as maximum, minimum, and average, have been developed for reducing the overall data traffic to save energy.

New Words

spatially	['speiʃəli]	adv.	空间地
autonomous	[ɔ:'tɔnəməs]	adj.	自治的
vibration	[vai'breiʃən]	n.	振动，颤动，摇动，摆动
pollutant	[pə'lu:tənt]	n.	污染物质
motivate	['məutiveit]	v.	推动，激发
battlefield	['bætlfi:ld]	n.	战场，沙场
surveillance	[sə:'veiləns]	n.	监视，监督
microcontroller	[,maikrəukən'trəulə]	n.	微控制器
interface	['intəfeis]	n.	界面连接，接口连接
		v.	连接
mote	[məut]	n.	尘埃，微粒
microscopic	[maikrə'skɔpik]	adj.	用显微镜可见的，精微的
corresponding	[,kɔris'pɔndiŋ]	adj.	相应的

---

① The Open Geospatial Consortium (OGC) is an international industry consortium ([kən'sɔ:tjəm]n. 社团，协会，联盟) of 438 companies, government agencies and universities participating in a consensus process to develop publicly available interface standards. OGC® Standards support interoperable solutions that "geo-enable" the Web, wireless and location-based services (基于位置的服务) and mainstream IT. The standards empower technology developers to make complex spatial information and services accessible and useful with all kinds of applications.

topology	[təˈpɔlədʒi]	n. 拓扑，布局
propagation	[ˌprɔpəˈgeiʃən]	n. (声波，电磁辐射等)传播
heterogeneity	[ˌhetərəudʒiˈniːiti]	n. 异种，异质，不同成分
withstand	[wiðˈstænd]	vt. 抵挡，经受住
harsh	[hɑːʃ]	adj. 粗糙的，荒芜的，苛刻的
unattended	[ˈʌnəˈtendid]	adj. 没人照顾的，未被注意的
inclusion	[inˈkluːʒən]	n. 包含，内含物
gateway	[ˈgeitwei]	n. 门，通路，网关
forward	[ˈfɔːwəd]	vt. 转发，传输
modem	[ˈməudəm]	n. 调制解调器
proprietary	[prəˈpraiətəri]	adj. 私有的，专有的 n. 所有者，所有权
predominant	[priˈdɔminənt]	adj. 卓越的，支配的，主要的，突出的，有影响的
inherent	[inˈhiərənt]	adj. 固有的，内在的，与生俱来的
scarce	[skɛəs]	adj. 缺乏的，不足的，稀有的，不充足的
hostile	[ˈhɔstail]	adj. 敌对的，敌方的
algorithm	[ˈælgəriðəm]	n. 算法
maximization	[ˌmæksəmaiˈzeiʃən]	n. 最大值化，极大值化
robustness	[rəˈbʌstnis]	n. 鲁棒性，稳健性，健壮性
debugging	[diːˈbʌgiŋ]	n. 调试
end-user	[ˈendˌjuːzə]	n. 终端用户
primitive	[ˈprimitiv]	adj. 简单的，粗糙的
property	[ˈprɔpəti]	n. 性质，特性
multithreading	[ˈmʌltiˈθrediŋ]	n. 多线程，多线索
semantics	[siˈmæntiks]	n. 语义学
kernel	[ˈkəːnl]	n. 核
abstraction	[æbˈstrækʃən]	n. 抽象
simulation	[ˌsimjuˈleiʃən]	n. 仿真，模拟
option	[ˈɔpʃən]	n. 选项，选择
paradigm	[ˈpærədaim]	n. 范例
flocking	[ˈflɔkiŋ]	n. 群集，成群结队而行
publish	[ˈpʌbliʃ]	v. 出版，刊印，公布，发表
notation	[nəuˈteiʃən]	n. 符号
quantify	[ˈkwɔntifai]	vt. 确定数量，量化
emergent	[iˈməːdʒənt]	adj. 紧急的，突然出现的，自然发生的
collapse	[kəˈlæps]	n. 倒塌，崩溃，失败 vi. 倒塌，崩溃，瓦解，失败
body	[ˈbɔdi]	n. 主体

allocate	[ˈæləukeit]	vt. 分派，分配
metadata	[ˈmetəˈdeitə]	n. 元数据
redundant	[riˈdʌndənt]	adj. 多余的，冗余的
directionality	[diˌrekʃənˈnæliti]	n. 方向性，定向性
aggregate	[ˈægrigeit]	v. 聚集，集合，合计
maximum	[ˈmæksiməm]	n. 最大量，最大限度，极大
		adj. 最高的，最多的，最大极限的
minimum	[ˈminiməm]	adj. 最小的，最低的
		n. 最小值，最小化

❖ Phrases

Wireless Sensor Network (WSN)	无线传感器网络
radio transceiver	无线电收发器
internal antenna	内置天线
electronic circuit	电子电路
energy source	能源
energy harvesting	能量收集，能量采集
a grain of dust	一粒灰尘
star network	星型网络
mesh network	网状网络
cope with	成功应付，妥善处理
processing unit	处理器，处理部件
communication device	通信设备
in the form of …	以……的形式
routing table	路由表
mobile phone network	移动电话网络
satellite phone	卫星电话
standardization body	标准化组织
Internet Engineering Task Force(IETF)	互联网工程任务组
International Society of Automation	美国国际自动化协会
fault tolerance	容错
virtual memory	虚拟内存
event-driven programming model	事件驱动编程模型
be composed of …	由……组成
event handler	事件句柄；事件处理程序
Agent-based Modeling and Simulation	基于代理的建模和仿真
mobile robot	移动式机器人，移动式遥控装置

in the vicinity of	在邻近
centralised architecture	集中式体系结构
distributed control architecture	分布控制式体系结构
be prone to	有……的倾向，易于
gather from …	从……获悉，从……收集
Open Geospatial Consortium (OGC)	开放地理空间联盟
be of no use	无用
information fusion	信息融合
making inference	推论
deal with	安排，处理，涉及

Abbreviations

ASIC (Application Specific Integrated Circuit)	特定用途集成电路
ARM (Advanced RISC Machines)	公司名
IEEE (Institute of Electrical and Electronics Engineers)	电气和电子工程师协会
MAC (Media Access Control)	介质访问控制，媒体存取控制
HART (Highway Addressable Remote Transducer Protocol)	可寻址远程传感器高速通道的开放通信协议
ISA (Industry Standard Architecture)	工业标准结构
P2P (Peer-to-Peer)	点对点技术

Exercises

I. Answer the following questions according to the text.

1. What does a Wireless Sensor Network (WSN) consist of?
2. What do sensor nodes consist of?
3. What do the base stations act as?
4. What do predominant standards commonly used in WSN communications include?
5. What is one major challenge in a WSN?
6. What are the two reasons Operating Systems for WSN more strongly resemble embedded systems?
7. How many ways are there to develop simulations of WSNs in general? What are they?
8. What are the reasons distributed control is used in WSNs?
9. What do some algorithms do to reduce communication costs?
10. What is information fusion called in WSNs? What has it been developed for?

II. Translate the following terms or phrases from English into Chinese and vice versa.

1. algorithm _____
2. robustness _____
3. communication device _____
4. multithreading _____
5. 处理器，处理部件 _____
6. forward _____
7. fault tolerance _____
8. 无线传感器网络 _____
9. topology _____
10. routing table _____

1. _____
2. _____
3. _____
4. _____
5. _____
6. _____
7. _____
8. _____
9. _____
10. _____

III. Fill in the blanks with the words given below.

| functionality | flash | fusion | ambient | humidity |
| transmission | sensing | node | energy | flexibility |

❖❖ Hardware Platform of WSN ❖❖

A WSN consists of spatially distributed sensor nodes. In a WSN, each sensor node is able to independently perform some processing and ___1___ tasks. Furthermore, sensor nodes communicate with each other in order to forward their sensed information to a central processing unit or conduct some local coordination such as data ___2___.

1. Embedded Processor

In a sensor node, the functionality of an embedded processor is to schedule tasks, process data and control the ___3___ of other hardware components. The types of embedded processors that can be used in a sensor node include Microcontroller, Digital Signal Processor (DSP), Field Programmable Gate Array (FPGA) and Application Specific Integrated Circuit (ASIC). Among all these alternatives, the Microcontroller has been the most used embedded processor for sensor nodes because of its ___4___ to connect to other devices and its cheap price. For example, the newest CC2531 development board provided by Chipcon (acquired by Texas Instruments) uses 8051 microcontroller, and the Mica2 Mote platform provided by Crossbow uses ATMega128L microcontroller.

2. Transceiver

A transceiver is responsible for the wireless communication of a sensor node. The various choices of wireless ___5___ media include Radio Frequency (RF), Laser and Infrared. RF based communication fits to most of WSN applications. The operational states of a transceiver are Transmit, Receive, Idle and Sleep.

3. Memory

Memories in a sensor node include in-chip ____6____ memory and RAM of a microcontroller and external flash memory.

4. Power Source

In a sensor node, power is consumed by sensing, communication and data processing. More ____7____ is required for data communication than for sensing and data processing. Power can be stored in batteries or capacitors. Batteries are the main source of power supply for sensor nodes.

To remove the energy constraint, some preliminary research working on energy-harvesting techniques for WSNs has also been conducted. Energy-harvesting techniques convert ____8____ ambient energy (e.g. solar, wind) to electrical energy and the aim is to revolutionize the power supply on sensor nodes.

5. Sensors

A sensor is a hardware device that produces a measurable response signal to a change in a physical condition such as temperature, pressure and ____9____. The continual analog signal sensed by the sensors is digitized by an analog-to-digital converter and sent to the embedded processor for further processing. Because a sensor ____10____ is a micro-electronic device powered by a limited power source, the attached sensors should also be small in size and consume extremely low energy. A sensor node can have one or several types of sensors integrated in or connected to the node.

IV. Translate the following passage from English to Chinese.

❖❖❖ WSN ❖❖❖

A Wireless Sensor Network (WSN) is a wireless network consisting of spatially distributed autonomous devices that use sensors to monitor physical or environmental conditions. These autonomous devices, or nodes, combine with routers and a gateway to create a typical WSN system. The distributed measurement nodes communicate wirelessly to a central gateway, which provides a connection to the wired world where you can collect, process, analyze, and present your measurement data. To extend distance and reliability in a wireless sensor network, you can use routers to gain an additional communication link between end nodes and the gateway.

National Instruments Wireless Sensor Networks offer reliable, low-power measurement nodes that operate for up to three years on 4 AA batteries and can be deployed for long-term, remote operation. The NI WSN protocol based on IEEE 802.15.4 and ZigBee technology provides a low-power communication standard that offers mesh routing capabilities to extend network distance and reliability. The wireless protocol you select for your network depends on your application requirements.

Text B

The Application of Wireless Sensor Network

1. Area monitoring

Area monitoring is a common application of WSNs. In area monitoring, the WSN is deployed over a region where some phenomenon is to be monitored. A military example is the use of sensors to detect enemy intrusion; a civilian example is the geo-fencing of gas or oil pipelines.

When the sensors detect the event being monitored (heat, pressure), the event is reported to one of the base stations, which then takes appropriate action (e.g., send a message on the internet or to a satellite). Similarly, wireless sensor networks can use a range of sensors to detect the presence of vehicles ranging from motorcycles to train cars.

2. Environmental sensing

The term Environmental Sensor Networks has evolved to cover many applications of WSNs to earth science research. This includes sensing volcanoes, oceans, glaciers, forests, etc.

3. Air pollution monitoring

Wireless sensor networks have been deployed in several cities to monitor the concentration of dangerous gases for citizens. These can take advantage of the ad-hoc wireless links rather than wired installations, which also make them more mobile for testing readings in different areas.

4. Forest fires detection

A network of Sensor Nodes can be installed in a forest to detect when a fire has started. The nodes can be equipped with sensors to measure temperature, humidity and gases which are produced by fires in the trees or vegetation. The early detection is crucial for a successful action of the firefighters; thanks to Wireless Sensor Networks, the fire brigade will be able to know when a fire is started and how it is spreading.

5. Greenhouse monitoring

Wireless sensor networks are also used to control the temperature and humidity levels inside commercial greenhouses. When the temperature and humidity drops below specific levels, the greenhouse manager must be notified via E-mail or cell phone text message, or host systems can trigger misting systems, open vents, turn on fans, or control a wide variety of system responses.

6. Landslide detection

A landslide detection system makes use of a wireless sensor network to detect the slight movements of soil and changes in various parameters that may occur before or during a landslide. And through the data gathered it may be possible to know the occurrence of landslides long before it actually happens.

7. Industrial monitoring

7.1 Machine health monitoring

WSNs have been developed for machinery Condition-Based Maintenance (CBM) as they offer significant cost savings and enable new functionalities. In wired systems, the installation of enough sensors is often limited by the cost of wiring. Previously inaccessible locations, rotating machinery, hazardous or restricted areas, and mobile assets can now be reached with wireless sensors.

7.2 Water/wastewater monitoring

There are many opportunities for using wireless sensor networks within the water/wastewater industries. Facilities not wired for power or data transmission can be monitored using industrial wireless I/O devices and sensors powered using solar panels or battery packs and also used in pollution control board.

7.3 Agriculture

Using WSNs within the agricultural industry is increasingly common; using a wireless network frees the farmer from the maintenance of wiring in a difficult environment. Gravity feed water systems can be monitored using pressure transmitters to monitor water tank levels, pumps can be controlled using wireless I/O devices and water use can be measured and wirelessly transmitted back to a central control center for billing. Irrigation automation enables more efficient water use and reduces waste.

7.4 Structural monitoring

Wireless sensors can be used to monitor the movement within buildings and infrastructure such as bridges, flyovers, embankments, tunnels, etc. enabling engineering practices to monitor assets remotely with out the need for costly site visits, as well as having the advantage of daily data, whereas traditionally this data was collected weekly or monthly, using physical site visits, involving either road or rail closure in some cases. It is also far more accurate than any visual inspection that would be carried out.

New Words

deploy	[di'plɔi]	v. 展开，配置，使用
phenomenon	[fi'nɔminən]	n. 现象
military	['militəri]	adj. 军事的，军方的，军队的
civilian	[si'viljən]	adj. 民用的，平民的，民间的 n. 平民

geo-fencing	[dʒiːəu-ˈfensiŋ]	n. 地理围栏
volcano	[ˈvəukeinəu]	n. 火山
glacier	[ˈglæsiə]	n. 冰山
concentration	[ˌkɔnsenˈtreiʃən]	n. 浓度
install	[inˈstɔːl]	vt. 安装,安置
humidity	[hjuːˈmiditi]	n. 湿度
vegetation	[vedʒiˈteiʃən]	n. 植物,草木
firefighter	[ˈfaiəfaitə]	n. 消防队员
commercial	[kəˈməʃəl]	adj. 商业的,商用的
greenhouse	[ˈgriːhaus]	n. 温室
notify	[ˈnəutifai]	v. 通报
trigger	[ˈtrigə]	vt. 引发,引起,触发
vent	[vent]	n. 通风口
landslide	[ˈlændslaid]	n. 山崩,山体滑坡
occurrence	[əˈkərəns]	n. 出现,发生
inaccessible	[inækˈsesibl]	adj. 不能进入的,不能达到的
rotating	[rəuˈteitiŋ]	adj. 旋转的
detection	[diˈtekʃən]	n. 察觉,发觉,探测,发现
hazardous	[ˈhæzədəs]	adj. 危险的,冒险的
wastewater	[ˈweistwɔːtə]	n. 废水
flyover	[ˈflaiəuvə]	n. 天桥,立交桥
embankment	[imˈbæŋkmənt]	n. 堤防,筑堤
tunnel	[ˈtʌnl]	n. 隧道,地道
closure	[ˈkləuʒə]	n. 关闭
		vt. 使终止
inspection	[inˈspekʃən]	n. 视察

Phrases

area monitoring	区域监测
enemy intrusion	敌人入侵
oil pipeline	输油管
base station	基站,基地
environmental sensing	环境感应
environmental sensor network	环境传感器网络
fire brigade	消防队
commercial greenhouses	商业性暖房,商业性大棚
misting system	喷雾系统
long before	很早以前

Condition-Based Maintenance (CBM)	状态维护，状态维修
solar panel	太阳电池板
battery pack	电池组
pollution control board	污染控制局
agricultural industry	农业
gravity feed water system	重力给水系统
water tank	水箱，水槽
irrigation automation	自动灌溉
physical site visit	实地考察

Exercises

I. Fill in the blanks with the information given in the text.

1. A military example of area monitoring is the use of sensors to _____; a civilian example is _____.

2. Wireless Sensor Networks can use a range of sensors to detect the presence of vehicles ranging from _____ to _____.

3. The term Environmental Sensor Networks has evolved to cover many applications of WSNs to earth science research. This includes sensing _____, _____, _____, _____ etc.

4. Thanks to Wireless Sensor Networks, the fire brigade will be able to know _____ and _____.

5. Wireless Sensor Networks are also used to control the temperature and humidity levels inside _____.

6. A landslide detection system makes use of a Wireless Sensor Network to detect _____ and changes in various parameters that may occur before or during _____.

7. Wireless Sensor Networks have been developed for machinery Condition-Based Maintenance (CBM) as they offer _____ and enable _____.

8. Facilities not wired for power or data transmission can be monitored using _____.

9. Gravity feed water systems can be monitored using pressure transmitters to _____. Irrigation automation enables _____ and reduces waste.

10. Wireless sensors can be used to monitor _____ and infrastructure such as bridges, flyovers, embankments, tunnels etc. It is also _____ than any visual inspection that would be carried out.

II. Translate the following terms or phrases from English into Chinese and vice versa.

1. deploy _____   1. _____

2. battery pack
3. 环境感应
4. geo-fencing
5. install
6. base station
7. concentration
8. humidity
9. 环境传感器网络
10. solar panel

2. ___
3. ___
4. ___
5. ___
6. ___
7. ___
8. ___
9. ___
10. ___

Text A 参考译文

无线传感器网络

一个无线传感器网络(WSN)由分布在不同空间的自治传感器组成,以便监控物质或环境条件,如温度、声音、振动、压力、运动或污染物质,并且协作地把数据传输到主要地点。更现代的网络是双向的,其也能控制传感器的行为。曾经,无线传感器网络的发展是由军事应用推动的,如战场监控。今天,这些网络已经用于许多工业领域和消费领域,如工业过程的监控、机器的健康监控等。

无线传感器网络由许多"节点"构成——从几个节点到几百节点甚至几千节点,每个节点都连接着一个(或有时数个)传感器。这样的传感器网络节点通常包含以下几部分:一个带有内置天线或可连接外部天线的无线电收发器、一个微型控制器、一个连接传感器和电源的电路、常用电池或内嵌的能量采集器。传感器节点的大小各异,从鞋盒大小到尘粒大小不等。传感器节点的成本也有多种,从几美元到几百美元不等,这取决于每个节点的复杂性。传感器节点的大小和成本决定其资源,如能量、内存、计算速度和通信带宽。无线传感器网络的拓扑也有多种,从简单的星型网络到高级的多跳跃无线掩膜网络。网络跳跃之间的传播技术在迅速发展。

在计算机科学和电信学中,无线传感器网络是众多研究机构积极研究的领域,也是每年会议的热点。

1. 特点

无线传感器网络的主要特点包括(如图 8.1 所示,图略):
• 能够对节点所用电池或能量采集器的能源消耗进行约束;
• 能够应付节点失效;
• 节点可迁移;
• 动态的网络拓扑;
• (可处理)通信故障;

- 大规模部署的可测量性；
- 能经受严酷环境的考验；
- 易用；
- 自主运行；
- 能源消耗(即节能)。

可以把传感器节点想像成小型计算机，其接口和部件非常简单。它们通常由以下几部分组成：计算能力有限的处理器和有限的内存、传感器或 MEMS(包括特定条件的电路)、通信设备(通常是无线电收发器或光收发器)以及一般为电池形式的电源，还可能包括能源采集模块、辅助的 ASIC 以及辅助的通信设备(如 RS-323 或 USB)。

基站带有一个或多个 WSN 部件，还带有许多可计算的、有能量的及通信的资源。它们通常把 WSN 上的数据转发到服务器，起传感器节点与终端用户之间的网关的作用。路由网络的其他特殊部件是路由器，是为计算、规划以及分配路由表而设计的。用于外部连接的技术有很多，包括手机网络、卫星电话、无线电调制解调器、远程 WiFi 链接等。许多基站基于 ARM 并运行于嵌入式 Linux 系统中。

2. 平台

2.1 标准和规范

用于无线传感器网络的几个流行标准要么已经认可，要么正在开发中。在 WSN 领域有许多标准化团体。IEEE 关注物理和 MAC 层。互联网工程任务组(IETF)从事第三层及以上的标准研究。除此之外，像美国国际自动化协会这样的团体提供垂直解决方案，覆盖了全部协议层。当然，还有几种非标准的、专有机制和规范。

用于 WSN 的标准比用于其他计算系统的少得多，这些标准使得大多数系统不能够在不同系统之间直接通信。WSN 通信中常用的标准包括：

- WirelessHART
- ISA100
- IEEE 1451
- ZigBee/802.15.4

2.2 硬件

WSN 中的一个主要挑战是生产低成本和微小的传感器节点。生产 WSN 硬件的小公司数量不断增加，商业境况与上世纪七十年代的家庭计算相仿。许多节点仍然处于研发阶段，其相关软件的研发更明显。同样，传感器网络一直采用非常低能量的数据收集方法。

2.3 软件

能量是 WSN 节点最匮乏的资源，并且它也决定了 WSN 的寿命。WSN 需要大量地部署在许多不同的环境中，包括远程和敌方区域，在那些地方特定通信尤为关键。因此，算法和协议要能解决以下问题：

- 寿命最大化；
- 健壮并容错；
- 自配置。

WSN 软件研究中的重要议题是：
- 操作系统；
- 安全性；
- 机动性；
- 可用性——用于部署和管理的人界面、调试和终端用户控制；
- 中间件——在高级软件和该系统之间的中级简要设计。

2.4 操作系统

用于无线传感器网络的操作系统比其他普通操作系统简单得多。它们更像嵌入式系统，有以下两个原因：首先，无线传感器网络典型地部署在特定的应用中，而不是通用平台；其次，低成本和低电能的需求使大多数无线传感器节点具有低功率微控制器，确保如虚拟内存这样的机制或者不需要，或者因太贵而不部署。

因此，可以在传感器网络中使用像 eCos 或 uC/OS 这样的嵌入式操作系统。但是，这样的操作系统通常需要具有实时特性。

TinyOS 或许是第一个专门为无线传感器网络设计的操作系统。TinyOS 基于事件驱动编程模式而不是多线程模式。TinyOS 程序由可运行结束的事件处理器沙漏任务组成。当一个外部事件(如输入数据包或传感器读入)发生时，TinyOS 给适当的事件处理器发送信号使其处理该事件。事件处理器稍后可以布置任务，这些任务由 TinyOS 内核安排。

LiteOS 是最新开发的无线传感器网络操作系统，它实质上类似于 UNIX 系统并支持 C 编程语言。

3. WSN 仿真

通常，可以使用两种方法来开发 WSN 仿真：或者使用定制平台来开发，或者开发自己的仿真平台。

3.1 仿真器

确切地说，目前基于代理的建模和仿真是唯一范式，它允许仿真无线传感器环境中非常复杂的行为(如群集)。

像 QualNet、NetSim 及 NS2 这样的仿真器可以用来仿真无线传感器网络。

3.2 WSN 基于代理的仿真

无线传感器及其特殊网络的基于代理的仿真是较新的范式。基于代理建模最初是基于社会模拟。IEEE 通信杂志最新发布的一篇基于代理仿真的论文给出了开发定制基于代理仿真模型的示例和方法，这些方法可以用于短期(几个小时)的无线传感器、移动式遥控装置和 P2P 网络；也提出了一个正式的基于代理仿真框架形式的规范，该规范使用 Z 符号表示采用基于代理仿真，以便提供对传感器环境中复杂行为的仿真。实践已经表明，基于代理仿真可以对 WSN 节点附近突发行为的定量分析进行建模和仿真。

4. 其他概念

4.1 分布式传感器网络

如果在传感器网络中使用集中式结构，并且当中心节点故障时，整个网络会崩溃。但是，使用分布式控制结构会增加传感器网络的可靠性。

在 WSN 中使用分布式控制的理由如下：
- 传感器节点容易出故障；
- 分布式控制可以更好地收集数据；
- 分布式控制可以在中心节点故障时提供备份节点。

即使使用了分布式控制，也没有一个中心实体来分配资源，它们必须实现自主组织。

4.2 数据集成和传感器网

从无线传感器网络中收集的数据通常以数字形式保存在中心基站上。另外，开放地理空间联盟为协同工作接口和元数据编码提供了具体规定标准，它们能把异种传感器网的数据实时整合到因特网中，让每个人都可以通过网络浏览器来监控无线传感器网络。

4.3 网内处理

为了减少通信成本，一些算法删除或减少了节点的冗余传感器信息并避免转发无用数据。因为节点可以检查它们所转发的数据，所以可以对数据进行测量，例如，测量来自其他节点数据的平均值或方向。

4.4 信息融合

在无线传感器网络中，信息融合(也叫做数据融合)的研制主要是为了可以通过对收集的数据进行过滤、聚集以及推测来处理传感器数据。信息融合综合了多种资源以改善信息：使信息更便宜、更有质量或更适用。在无线传感器网络领域中，为了通过减少数据流量来节省能量，如最大值、最小值、平均值这样的简单合并计算技术已经开发出来了。

Unit 9 Network

Text A

IEEE 802.15.4

IEEE 802.15.4 is a standard which specifies the physical layer[①] and media access control for low-rate wireless personal area networks (LR-WPANs). It is maintained by the IEEE 802.15 working group. It is the basis for the ZigBee, ISA100.11a, Wireless HART, and MiWi[②] specifications, each of which further extends the standard by developing the upper layers which are not defined by 802.15.4. Alternatively, it can be used with 6LoWPAN[③] and standard Internet protocols to build a Wireless Embedded Internet (See Figure 9.1).

1. Overview

IEEE standard 802.15.4 intends to offer the fundamental lower network layers of a type of Wireless Personal Area Network (WPAN) which focuses on

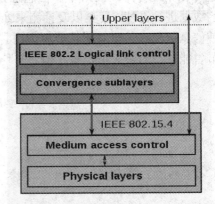

Figure 9.1 IEEE 802.15.4 protocol stack

low-cost, low-speed ubiquitous communication between devices (in contrast with other, more

---

① The physical layer is the first and lowest layer in the seven-layer OSI model of computer networking. The physical layer consists of the basic networking hardware transmission technologies of a network. It is a fundamental layer underlying the logical data structures of the higher level functions in a network.

② MiWi and MiWi P2P (Peer-to-Peer, 对等, 点对点) are proprietary wireless protocols designed by Microchip Technology that uses small, low-power digital radios based on the IEEE 802.15.4 standard for Wireless Personal Area Networks (WPANs). It is designed for low data transmission rates and short distance, cost constrained networks, such as industrial monitoring and control, home and building automation, remote control, low-power wireless sensors, lighting control and automated meter reading.

③ 6LoWPAN is an acronym of IPv6 over Low power Wireless Personal Area Networks. The 6LoWPAN concept originated from the idea that "the Internet Protocol could and should be applied even to the smallest devices," and that low-power devices with limited processing capabilities should be able to participate in (参与, 分享) the Internet of Things.

end-user oriented approaches, such as WiFi). The emphasis is on very low cost communication of nearby devices with little to no underlying infrastructure, intending to exploit this to lower power consumption even more.

The basic framework conceives a 10-meter communications range with a transfer rate[①] of 250 kbit/s. Tradeoffs are possible to favor more radically embedded devices[②] with even lower power requirements, through the definition of not one, but several physical layers. Lower transfer rates of 20 and 40 kb/s were initially defined, with the 100 kb/s rate being added in the current revision.

Even lower rates can be considered with the resulting effect on power consumption. As already mentioned, the main identifying feature of 802.15.4 among WPAN's is the importance of achieving extremely low manufacturing and operation costs and technological simplicity, without sacrificing flexibility or generality.

Important features include real-time suitability by reservation of guaranteed time slots, collision avoidance through CSMA/CA[③] and integrated support for secure communications. Devices also include power management functions such as link quality and energy detection.

802.15.4-conformant devices may use one of three possible frequency bands for operation.

2. Protocol architecture

Devices are conceived to interact with each other over a conceptually simple wireless network. The definition of the network layers is based on the OSI model[④]; although only the lower layers are defined in the standard, interaction with upper layers is intended, possibly using

---

① In telecommunications and computing, bit rate (sometimes written bitrate, data rate or as a variable R) is the number of bits that are <u>conveyed</u> ([kən'veɪ]vt. 传送，传达) or processed per unit of time.

② An embedded system is a computer system designed for specific control functions within a larger system, often with real-time computing constraints. It is embedded as part of a complete device often including hardware and mechanical parts. By contrast, a <u>general-purpose computer</u> (通用计算机), such as a Personal Computer (PC), is designed to be flexible and to meet a wide range of end-user needs.

③ Carrier Sense Multiple Access with Collision Avoidance (CSMA/CA), in computer networking, is a wireless network multiple access method in which: (1) a carrier sensing scheme is used; (2) a node wishing to transmit data has to first listen to the channel for a predetermined amount of time to determine whether or not another node is transmitting on the channel within the wireless range. If the channel is sensed "idle", then the node is permitted to begin the transmission process. If the channel is sensed as "busy", the node <u>defers</u> ([dɪ'fɜː] vi. 推迟，延期) its transmission for a random period of time. Once the transmission process begins, it is still possible for the actual transmission of application data to not occur.

④ The Open Systems Interconnection (OSI) model is a product of the Open Systems Interconnection effort at the International Organization for Standardization. It is a prescription of characterizing and standardizing the functions of a communications system <u>in terms of</u> (根据，按照，在……方面) abstraction layers. Similar communication functions are grouped into logical layers. A layer serves the layer above it and is served by the layer below it.

a IEEE 802.2 logical link control[①] sublayer accessing the Media Access Control(MAC) through a convergence sublayer. Implementations may rely on external devices or be purely embedded, self-functioning devices.

2.1 The Physical Layer

The Physical Layer (PHY) ultimately provides the data transmission service, as well as the interface to the physical layer management entity, which offers access to every layer management function and maintains a database of information on related Personal Area Networks(PAN). Thus, the PHY manages the physical RF transceiver and performs channel selection and energy and signal management functions. It operates on one of three possible unlicensed frequency bands:

- 868.0–868.6 MHz: Europe, allows one communication channel (2003, 2006);
- 902–928 MHz: North America, up to ten channels (2003), extended to thirty (2006);
- 2400–2483.5 MHz: worldwide use, up to sixteen channels (2003, 2006).

The original 2003 version of the standard specifies two physical layers based on Direct Sequence Spread Spectrum (DSSS)[②] techniques: one working in the 868/915 MHz bands with transfer rates of 20 and 40 kbit/s, and one in the 2450 MHz band with a rate of 250 kbit/s.

The 2006 revision improves the maximum data rates of the 868/915 MHz bands, bringing them up to support 100 and 250 kb/s as well. Moreover, it goes on to define four physical layers depending on the modulation method used. Three of them preserve the DSSS approach: in the 868/915 MHz bands, using either binary or offset quadrature phase shift keying (the second of which is optional); in the 2450 MHz band, using the latter. An alternative, optional 868/915 MHz layer is defined using a combination of binary keying and amplitude shift keying[③] (thus based on Parallel, not Sequential Spread Spectrum, PSSS). Dynamic switching between supported 868/915 MHz PHYs is possible.

Beyond these three bands, the IEEE 802.15.4c study group is considering the newly opened 314–316 MHz, 430–434 MHz, and 779–787 MHz bands in China, while the IEEE 802.15 Task Group 4d is defining an amendment to the existing standard 802.15.4-2006 to support the new 950 MHz–956 MHz band in Japan. First standard amendments by these groups were released in

---

① The Logical Link Control (LLC) data communication protocol layer is the upper sub-layer of the data link layer (which is itself layer 2, just above the physical layer) in the seven-layer OSI reference model. It provides multiplexing (['mʌltipleksiŋ]n. 多路技术) mechanisms that make it possible for several network protocols (IP, IPX, Decnet and Appletalk) to coexist ([kəuig'zist]vi. 共存) within a multipoint (['mʌltipɔint]adj. 多点(式)的，多位置的) network and to be transported over the same network media, and can also provide flow control and Automatic Repeat Request (ARQ，自动重复请求) error management mechanisms.

② In telecommunications, Direct-Sequence Spread Spectrum (DSSS) is a modulation technique. As with other spread spectrum technologies, the transmitted signal takes up more bandwidth than the information signal that is being modulated. The name "spread spectrum" comes from the fact that the carrier signals occur over the full bandwidth (spectrum) of a device's transmitting frequency. Certain IEEE 802.11 standards use DSSS signaling (['signəliŋ]n. 发信号).

③ Amplitude-Shift Keying (ASK) is a form of modulation that represents digital data as variations in the amplitude (['æmplitjuːd]n. 振幅) of a carrier wave (载波).

April 2009.

In August 2007, IEEE 802.15.4a was released expanding the four PHYs available in the earlier 2006 version to six, including one PHY using Direct Sequence Ultra-WideBand (UWB)[①] and another using Chirp Spread Spectrum (CSS)[②]. The UWB PHY is allocated frequencies in three ranges: below 1 GHz, between 3 and 5 GHz, and between 6 and 10 GHz. The CSS PHY is allocated spectrum in the 2450 MHz band.

In April 2009, IEEE 802.15.4c and IEEE 802.15.4d were released expanding the available PHYs with several additional PHYs: one for 780 MHz band using O-QPSK or MPSK, another for 950 MHz using GFSK or BPSK.

2.2 The MAC layer

The Medium Access Control (MAC) enables the transmission of MAC frames through the use of the physical channel. Besides the data service, it offers a management interface and itself manages access to the physical channel and network beaconing[③]. It also controls frame validation, guarantees time slots and handles node associations. Finally, it offers hook points for secure services.

2.3 Higher layers

Other higher-level layers and interoperability sublayers are not defined in the standard. There exist specifications, such as ZigBee, which build on this standard to propose integral solutions.

3. Network model

3.1 Node types

The standard defines two types of network node.

The first one is the Full-Function Device (FFD). It can serve as the coordinator of a personal area network just as it may function as a common node. It implements a general model of communication which allows it to talk to any other device: it may also relay messages, in which case it is dubbed a coordinator.

On the other hand there are Reduced-Function Devices (RFD). These are meant to be extremely simple devices with very modest resource and communication requirements; due to

---

① Ultra-WideBand (aka UWB, ultra-wide band, ultraband, etc.) is a radio technology pioneered by Robert A. Scholtz and others that can be used at very low energy levels for short-range high-bandwidth (短距离、高带宽) communications by using a large portion of the radio spectrum (射频频谱). UWB has traditional applications in non-cooperative radar imaging (雷达图像). Most recent applications target sensor data collection, precision locating and tracking applications.

② In digital communications, Chirp Spread Spectrum (CSS) is a spread spectrum technique that uses wideband linear frequency modulated chirp pulses to encode information. A chirp is a sinusoidal ([ˌsaɪnəˈsɔɪdəl]adj.正弦曲线) signal whose frequency increases or decreases over a certain amount of time.

③ The process that allows a network to self-repair (自修复) networks problems. The stations on the network notify the other stations on the ring when they are not receiving the transmissions. Beaconing is used in Token ring and FDDI networks.

this, they can only communicate with FFD's and can never act as coordinators.

3.2 Topologies

Networks can be built as either Peer-to-Peer① or star networks(See Figure 9.2). However, every network needs at least one FFD to work as the coordinator of the network. Networks are thus formed by groups of devices separated by suitable distances. Each device has a unique 64-bit identifier, and if some conditions are met short 16-bit identifiers can be used within a restricted environment. Namely, within each PAN domain, communications will probably use short identifiers.

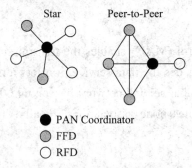

Figure 9.2   IEEE 802.15.4 star and Peer-to-Peer

Peer-to-Peer (or Point-to-Point) networks can form arbitrary patterns of connections, and their extension is only limited by the distance between each pair of nodes. They are meant to serve as the basis for ad hoc networks② capable of performing self-management and organization. Since the standard does not define a network layer, routing is not directly supported, but such an additional layer can add support for multihop communications. Further topological restrictions may be added; the standard mentions the cluster tree(See Figure 9.3) as a structure which exploits the fact that an RFD may only be associated with one FFD at a time to form a network where RFD's are exclusively leaves of a tree, and most of the nodes are FFD's. The structure can be extended as a generic mesh network③ whose nodes are cluster tree networks with a local coordinator for each cluster, in addition to the global coordinator.

---

① Peer-to-Peer (abbreviated to P2P) refers to a computer network in which each computer in the network can act as a client or server for the other computers in the network, allowing shared access to files and peripherals without the need for a central server. P2P networks can be used for sharing content such as audio, video, data or anything in digital format.

② A wireless ad-hoc network is a <u>decentralized</u> ([diːˈsentrəlaiz]n. 分散) type of wireless network. The network is ad hoc because it does not rely on a <u>preexisting</u> ([ˈpriːigˈzist]v. 先前存在) infrastructure, such as routers in wired networks or access points in managed (infrastructure) wireless networks. Instead, each node participates in routing by forwarding data for other nodes, and so the determination of which nodes forward data is made dynamically based on the network connectivity. In addition to the classic routing, ad hoc networks can use flooding for forwarding the data.

③ Mesh networking (topology) is a type of networking where each node must not only capture and <u>disseminate</u> ([diˈsemineit]v. 传播) its own data, but also serve as a relay for other nodes, that is, it must collaborate to <u>propagate</u> ([ˈprɔpəgeit]v. 传播) the data in the network.

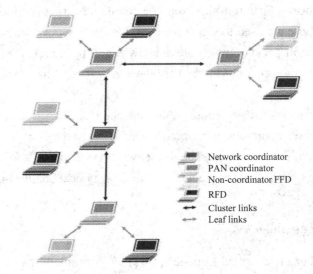

Figure 9.3   IEEE 802.15.4 cluster tree

A more structured star pattern is also supported, where the coordinator of the network will necessarily be the central node. Such a network can originate when an FFD decides to create its own PAN and declare itself its coordinator, after choosing a unique PAN identifier. After that, other devices can join the network, which is fully independent from all other star networks.

4. Data transport architecture

Frames are the basic unit of data transport, of which there are four fundamental types (data, acknowledgment, beacon and MAC command frames), which provide a reasonable tradeoff between simplicity and robustness. Additionally, a superframe structure, defined by the coordinator, may be used, in which case two beacons act as its limits and provide synchronization to other devices as well as configuration information. A superframe consists of sixteen equal-length slots, which can be further divided into an active part and an inactive part, during which the coordinator may enter power saving mode, not needing to control its network.

Within superframes contention occurs between their limits, and is resolved by CSMA/CA[①]. Every transmission must end before the arrival of the second beacon. As mentioned before, applications with well-defined bandwidth needs can use up to seven domains of one or more contentionless guaranteed time slots, trailing at the end of the superframe. The first part of the superframe must be sufficient to give service to the network structure and its devices. Superframes are typically utilized within the context of low-latency devices, whose associations must be kept even if inactive for long periods of time.

Data transfers to the coordinator require a beacon synchronization phase, if applicable, followed by CSMA/CA transmission (by means of slots if superframes are in use);

---

① Short for Carrier Sense Multiple Access/Collision Detection, a set of rules determining how network devices respond when two devices attempt to use a data channel simultaneously (called a collision).

acknowledgment is optional. Data transfers from the coordinator usually follow device requests: if beacons are in use, these are used to signal requests; the coordinator acknowledges the request and then sends the data in packets which are acknowledged by the device. The same is done when superframes are not in use, only in this case there are no beacons to keep track of pending messages.

Point-to-Point networks may either use unslotted CSMA/CA or synchronization mechanisms; in this case, communication between any two devices is possible, whereas in "structured" modes one of the devices must be the network coordinator.

In general, all implemented procedures follow a typical request-confirm/indication-response classification.

5. Reliability and security

The physical medium is accessed through a CSMA/CA protocol. Networks which are not using beaconing mechanisms utilize an unslotted variation which is based on the listening of the medium, leveraged by a random exponential backoff[①] algorithm; acknowledgments do not adhere to this discipline. Common data transmission utilizes unallocated slots when beaconing is in use; again, confirmations do not follow the same process.

Confirmation messages may be optional under certain circumstances. Whatever the case, if a device is unable to process a frame at a given time, it simply does not confirm its reception: timeout-based retransmission can be performed a number of times, following after that a decision of whether to abort or keep trying.

Because the predicted environment of these devices demands maximization of battery life, the protocols tend to favor the methods which lead to it, implementing periodic checks for pending messages, the frequency of which depends on application needs.

Regarding secure communications, the MAC sublayer offers facilities which can be harnessed by upper layers to achieve the desired level of security. Higher-layer processes may specify keys to perform symmetric cryptography to protect the payload and restrict it to a group of devices or just a Point-to-Point link; these groups of devices can be specified in access control lists. Furthermore, MAC computes freshness checks between successive receptions to ensure that presumably old frames, or data which is no longer considered valid, does not transcend to higher layers.

In addition to this secure mode, there is another, insecure MAC mode, which allows access control lists merely as a means to decide on the acceptance of frames according to their (presumed) source.

---

① Exponential backoff is an algorithm that uses feedback (['fi:dbæk]n. 反馈) to multiplicatively decrease the rate of some process, in order to gradually (['grædjuəli]adv. 逐渐地) find an acceptable rate. In a variety of computer networks, binary exponential backoff or truncated binary exponential backoff (截断二进制指数后移) refers to an algorithm used to space out repeated retransmissions of the same block of data (数据块), often as part of network congestion ([kən'dʒestʃən]n. 拥塞) avoidance.

New Words

fundamental	[ˌfʌndəˈmentl]	adj. 基础的，基本的
		n. 基本原则，基本原理
emphasis	[ˈemfəsis]	n. 强调，重点
exploit	[iksˈplɔit]	v. 使用
conceive	[kənˈsiːv]	vt. 构思，以为，持有
		vi. 考虑，设想
tradeoff	[treidɔf]	n. 折衷，权衡，(公平)交易
radically	[ˈrædikəli]	adv. 根本上
revision	[riˈviʒən]	n. 修订，修改，修正，修订本
sacrifice	[ˈsækrifais]	n. & v. 牺牲
generality	[ˌdʒenəˈræliti]	n. 一般性，普遍性，大部分
suitability	[ˌsjuːtəˈbiləti]	n. 合适，适当，相配，适宜性
guarantee	[ˌgærənˈtiː]	vt. 保证，担保
		n. 保证，保证书，担保，抵押品
conform	[kənˈfɔːm]	vt. 使一致，使遵守，使顺从
		vi. 符合，相似，适应环境
		adj. 一致的，顺从的
sublayer	[ˈsʌbˈleiə]	n. 子层，下层，次层
convergence	[kənˈvəːdʒəns]	n. 汇聚，集中，集合
implementation	[ˌimplimenˈteiʃən]	n. 执行
unlicensed	[ˈʌnˈlaisənst]	adj. 没有执照的，未经当局许可的，无节制的
amendment	[əˈmendmənt]	n. 改善，改正
spectrum	[ˈspektrəm]	n. 频谱，光谱，型谱
beaconing	[ˈbiːkəniŋ]	n. 信标
validation	[væliˈdeiʃən]	n. 确认
coordinator	[kəuˈɔːdineitə]	n. 协调者
relay	[ˈriːlei]	v. 转播
		n. 继电器
modest	[ˈmɔdist]	adj. 适度的
restricted	[risˈtriktid]	adj. 受限制的，有限的
multihop	[ˈmʌltihɔp]	n. 多次反射
restriction	[risˈtrikʃən]	n. 限制，约束
acknowledgment	[əkˈnɔlidʒmənt]	n. 应答，承认
superframe	[ˈsjuːpəˌfreim]	n. 超帧
contention	[kənˈtenʃən]	n. 争用，争夺

trail	[treil]	vt. 跟踪，追踪，拉，拖
		n. 踪迹，痕迹，形迹
assumption	[əˈsʌmʃən]	n. 假定，设想，担任，承当
abort	[əˈbɔːt]	vi. 异常中断，中途失败
harness	[ˈhɑːnis]	n. 系在身上的绳子
		vt. 上马具，披上甲胄
payload	[ˈpeiˌləud]	n. 有效载荷
successive	[səkˈsesiv]	adj. 继承的，连续的
presumably	[priˈzjuːməbəli]	adv. 推测起来，大概
valid	[ˈvælid]	adj. 有效的，有根据的，正确的

Phrases

physical layer	物理层
Media Access Control(MAC)	介质访问控制
Personal Area Network(PAN)	个人局域网
Wireless Embedded Internet	无线嵌入式因特网
in contrast with	和……形成对比，和……形成对照
time slot	时间空档，时隙
collision avoidance	防止空中相撞
power management	动力管理
Direct Sequence Spread Spectrum (DSSS)	直接序列扩频，直接序列扩频通信
transfer rate	传输率
offset quadrature phase shift keying	偏移四相移相键控
binary keying	二进制键控
amplitude shift keying	幅移键控，幅变调制，振幅偏移键控法
Parallel，not Sequential Spread Spectrum (PSSS)	平行，而不是序列扩展频谱
study group	学习研讨会，研究小组
Direct Sequence Ultra-WideBand (UWB)	直接序列超宽带
Chirp Spread Spectrum (CSS)	线性调频技术
hook point	钩入点
Full-Function Device (FFD)	全功能设备，全功能装置
Reduced-Function Device (RFD)	功能简化设备
be meant to	有意要，打算
a pair of	一对
Ad hoc Network	自组织网络
cluster tree	簇状结构，簇状布局
give service to …	为……服务

low-latency device	低延迟的装置
by means of	依靠，用，借助于
pending message	待审消息
random exponential backoff	随机指数回退
under certain circumstance	在某种情况下
timeout-based retransmission	基于超时重传，基于超时重发
periodic check	定期检查
symmetric cryptography	对称加密

Abbreviations

MiWi	Microchip 无线协议
O-QPSK (Offset Quadrature Phase Shift Keying)	偏移四相相移键控，偏移正交相移键控
MPSK (Mary Phase Shift Keying)	多进制相移键控
GFSK (Gaussian Frequency Shift Keying)	高斯移频键控
BPSK (Binary Phase Shift Keying)	二进制移相键控
CSMA/CA (Carrier Sense Multiple Access with Collision Avoidance)	载波侦听多点访问/避免冲突

Exercises

I. Answer the following questions according to the text.

1. What is IEEE 802.15.4?
2. What does IEEE standard 802.15.4 intend to offer?
3. What are the possible unlicensed frequency bands the PHY operates on?
4. How does the Medium Access Control (MAC) enable the transmission of MAC frames?
5. How many types of network node does the standard define? What are they?
6. What can networks built as? What can Peer-to-Peer (or Point-to-Point) networks form?
7. What are frames?
8. What does a superframe consist of?
9. What happens if a device is unable to process a frame at a given time?
10. What does the MAC sublayer offer regarding secure communications?

II. Translate the following terms or phrases from English into Chinese and vice versa.

1. binary keying _____
2. superframe _____
3. 个人局域网 _____
4. Media Access Control _____
5. offset quadrature phase shift keying _____

1. _____
2. _____
3. _____
4. _____
5. _____

6. multihop _____   6. _____
7. 传输率 _____   7. _____
8. spectrum _____   8. _____
9. symmetric cryptography _____   9. _____
10. beaconing _____   10. _____

III. Fill in the blanks with the words given below.

| wireless | lock | unauthorized | seamless | specialized |
| server | findings | interconnecting | computing | physical |

◆◇ **WPAN** ◇◆

A Wireless Personal Area Network (WPAN) is a Personal Area Network—a network for __1__ devices centered around an individual person's workspace—in which the connections are __2__. Wireless PAN is based on the standard IEEE 802.15. The three kinds of wireless technologies used for WPAN are Bluetooth, Infrared Data Association, and WiFi.

A WPAN could serve to interconnect all the ordinary __3__ and communicating devices that many people have on their desk or carry with them today—or it could serve a more __4__ purpose such as allowing the surgeon and other team members to communicate during an operation.

A key concept in WPAN technology is known as "plugging in". In the ideal scenario, when any two WPAN-equipped devices come into close proximity (within several meters of each other) or within a few kilometers of a central __5__, they can communicate as if connected by a cable. Another important feature is the ability of each device to __6__ out other devices selectively, preventing needless interference or __7__ access to information.

The technology for WPANs is in its infancy and is undergoing rapid development. Proposed operating frequencies are around 2.4 GHz in digital modes. The objective is to facilitate __8__ operation among home or business devices and systems. Every device in a WPAN will be able to plug in to any other device in the same WPAN, provided they are within __9__ range of one another. In addition, WPANs worldwide will be interconnected. Thus, for example, an archeologist on site in Greece might use a PDA to directly access databases at the University of Minnesota in Minneapolis, and to transmit __10__ to that database.

IV. Translate the following passage from English to Chinese.

◆◇ **PAN** ◇◆

A Personal Area Network (PAN) is a computer network used for communication among computerized devices, including telephones and personal digital assistants. PANs can be used for communication among the personal devices themselves (intrapersonal communication), or for connecting to a higher level network and the Internet (an uplink). A wireless personal area network (WPAN) is a PAN carried over wireless network technologies such as IrDA, Bluetooth, Wireless USB, Z-Wave, ZigBee, or even Body Area Network. The reach of a WPAN varies from a few centimeters to a few meters. A PAN may also be carried over wired computer buses such as USB and FireWire.

ZigBee

ZigBee is a specification for a suite of high level communication protocols using small, low-power digital radios based on an IEEE 802 standard for personal area networks. Applications include wireless light switches, electrical meters with in-home-displays, and other consumer and industrial equipment that requires short-range wireless transfer of data at relatively low rates. The technology defined by the ZigBee specification is intended to be simpler and less expensive than other WPANs, such as Bluetooth. ZigBee is targeted at Radio-Frequency (RF) applications that require a low data rate, long battery life, and secure networking. ZigBee has a defined rate of 250 kbps best suited for periodic or intermittent data or a single signal transmission from a sensor or input device (See Figure 9.4).

The € 1 coin, shown for size reference, is about 23 mm (0.9 inch) in diameter.

Firgure 9.4   ZigBee module

The name refers to the waggle dance of honey bees after their return to the beehive.

1. Technical overview

ZigBee is a low-cost, low-power, wireless mesh network standard. The low cost allows the technology to be widely deployed in wireless control and monitoring applications. Low power-usage allows longer life with smaller batteries. Mesh networking provides high reliability and more extensive range. The technology is intended to be simpler and less expensive than other WPANs such as Bluetooth. ZigBee chip vendors typically sell integrated radios and microcontrollers with between 60 KB and 256 KB flash memory.

ZigBee operates in the Industrial, Scientific and Medical (ISM[①]) radio bands; 868 MHz in Europe, 915 MHz in the USA and Australia, and 2.4 GHz in most jurisdictions worldwide. Data transmission rates vary from 20 to 250 kb/s.

The ZigBee network layer natively supports both star and tree typical networks, and generic mesh networks. Every network must have one coordinator device, tasked with its creation, the control of its parameters and basic maintenance. Within star networks, the coordinator must be the central node. Both trees and meshes allows the use of ZigBee routers[②] to extend communication at the network level.

ZigBee builds upon the physical layer and medium access control defined in IEEE standard 802.15.4 (2003 version) for low-rate WPAN's. The specification goes on to complete the standard by adding four main components: network layer, application layer, ZigBee Device Objects (ZDO's) and manufacturer-defined application objects which allow for customization and favor total integration (See Figure 9.5).

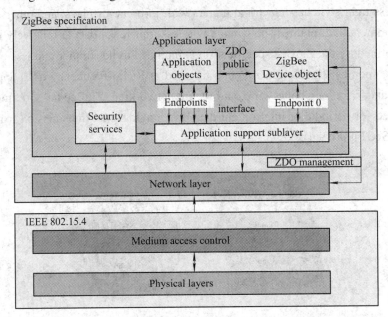

Figure 9.5 ZigBee protocol stack

Besides adding two high-level network layers to the underlying structure, the most significant improvement is the introduction of ZDO's. These are responsible for a number of tasks, which include keeping of device roles, management of requests to join a network, device discovery and security.

---

① The Industrial, Scientific and Medical (ISM) radio bands are radio bands (<u>portions</u> (portion[ˈpɔːʃən]n. 一部分) of the radio spectrum) reserved internationally for the use of Radio Frequency (RF) energy for industrial, scientific and medical purposes other than communications.

② Routing is the process of selecting paths in a network along which to send network traffic. Routing is performed for many kinds of networks, including the telephone network (<u>circuit switching</u> (线路交换)), electronic data networks (such as the Internet), and transportation networks.

ZigBee is not intended to support powerline networking[①] but to interface with it at least for smart metering[②] and smart appliance purposes.

Because ZigBee nodes can go from sleep to active mode in 30 msec or less, the latency can be low and devices can be responsive, particularly compared to Bluetooth wake-up delays, which are typically around three seconds. Because ZigBee nodes can sleep most of the time, average power consumption can be low, resulting in long battery life.

2. Uses

ZigBee protocols are intended for embedded applications requiring low data rates and low power consumption. The resulting network will use very small amounts of power — individual devices must have a battery life of at least two years to pass ZigBee certification.

Typical application areas include:

• Home Entertainment and Control—Home automation, smart lighting, advanced temperature control, safety and security, movies and music;

• Wireless Sensor Networks' — Starting with individual sensors like Telosb/Tmote and Iris from Memsic;

• Industrial control;

• Embedded sensing;

• Medical data collection;

• Smoke and intruder warning;

• Building automation.

3. Device types

There are three different types of ZigBee devices:

• ZigBee Coordinator (ZC): The most capable device, the coordinator forms the root of the network tree and might bridge to other networks. There is exactly one ZigBee Coordinator in each network since it is the device that started the network originally. It is able to store information about the network, including acting as the Trust Center & repository for security keys.

• ZigBee Router (ZR): As well as running an application function, a router can act as an intermediate router, passing on data from other devices.

---

① Power Line Communication or Power Line Carrier (PLC), also known as Power Line Digital Subscriber line (PLDS, 电力电路数字用户线路), mains communication, Power Line Telecom (PLT, 电力电路电信), Power Line Networking (PLN), or Broadband over Power Lines (BPL) are systems for carrying data on a conductor also used for electric power transmission (输电).

② A smart meter (智能仪表) is usually an electrical meter that records consumption of electric energy in intervals of an hour or less and communicates that information at least daily back to the utility for monitoring and billing purposes. Smart meters enable two-way communication between the meter and the central system. Unlike home energy monitors, smart meters can gather data for remote reporting.

• ZigBee End Device (ZED): Contains just enough functionality to talk to the parent node (either the coordinator or a router); it cannot relay data from other devices. This relationship allows the node to be asleep a significant amount of the time thereby giving long battery life. A ZED requires the least amount of memory, and therefore can be less expensive to manufacture than a ZR or ZC.

4. Protocols

The protocols build on recent algorithmic research (Ad-hoc On-demand Distance Vector, neuRFon[①]) to automatically construct a low-speed ad-hoc network of nodes. In most large network instances, the network will be a cluster of clusters. It can also form a mesh or a single cluster. The current ZigBee protocols support beacon and non-beacon enabled networks.

In non-beacon-enabled networks, an unslotted CSMA/CA channel access mechanism is used. In this type of network, ZigBee Routers typically have their receivers continuously active, requiring a more robust power supply. However, this allows for heterogeneous networks in which some devices receive continuously, while others only transmit when an external stimulus is detected. The typical example of a heterogeneous network is a wireless light switch: The ZigBee node at the lamp may receive constantly, since it is connected to the mains supply, while a battery-powered light switch would remain asleep until the switch is thrown. The switch then wakes up, sends a command to the lamp, receives an acknowledgment, and returns to sleep. In such a network the lamp node will be at least a ZigBee Router, if not the ZigBee Coordinator; the switch node is typically a ZigBee End Device.

In beacon-enabled networks, the special network nodes called ZigBee Routers transmit periodic beacons to confirm their presence to other network nodes. Nodes may sleep between beacons, thus lowering their duty cycle[②] and extending their battery life. Beacon intervals depend on data rate; they may range from 15.36 milliseconds to 251.65824 seconds at 250 kb/s, from 24 milliseconds to 393.216 seconds at 40 kb/s and from 48 milliseconds to 786.432 seconds at 20 kb/s. However, low duty cycle operation with long beacon intervals requires precise timing, which can conflict with the need for low product cost.

In general, the ZigBee protocols minimize the time the radio is on, so as to reduce power use. In beaconing networks, nodes only need to be active while a beacon is being transmitted. In non-beacon-enabled networks, power consumption is decidedly asymmetrical: some devices are always active, while others spend most of their time sleeping.

Except for the Smart Energy Profile 2.0, ZigBee devices are required to conform to the IEEE 802.15.4-2003 Low-Rate Wireless Personal Area Network (LR-WPAN) standard. The

---

① The neuRFon project (named for a combination of "neuron (['njuərɔn]n. 神经细胞, 神经元)" and "RF") was a research program begun in 1999 at Motorola Labs to develop ad hoc wireless networking for wireless sensor network applications.

② A duty cycle is the time that an entity spends in an active state (活动状态) as a fraction of (一小部分) the total time under consideration.

standard specifies the lower protocol layers—the Physical Layer (PHY), and the Media Access Control portion of the Data Link Layer (DLL). The basic channel access mode is "Carrier Sense, Multiple Access/Collision Avoidance" (CSMA/CA). That is, the nodes talk in the same way that people converse; they briefly check to see that no one is talking before they start. There are three notable exceptions to the use of CSMA. Beacons are sent on a fixed timing schedule, and do not use CSMA. Message acknowledgments also do not use CSMA. Finally, devices in Beacon Oriented networks that have low latency real-time requirements may also use Guaranteed Time Slots (GTS), which by definition do not use CSMA.

New Words

target	['tɑːgit]	n. 目标
periodic	[piəri'ɔdik]	adj. 周期的，定期的
intermittent	[ˌintə(ː)'mitənt]	adj. 间歇的，断断续续的
waggle	['wægl]	v. 来回摇动，摆动
beehive	['biːhaiv]	n. 蜂窝，蜂箱
extensive	[iks'tensiv]	adj. 广大的，广阔的，广泛的
jurisdiction	[ˌdʒuəris'dikʃən]	n. 管辖权，权限
generic	[dʒi'nerik]	adj. 一般的，普通的
favor	['feivə]	vt. 支持，赞成，促成
latency	['leitənsi]	n. 反应时间
entertainment	[entə'teinmənt]	n. 娱乐
intruder	[in'truːdə]	n. 入侵者
repository	[ri'pɔzitəri]	n. 储藏室，仓库
functionality	[ˌfʌŋkʃə'næliti]	n. 功能性，泛函性
vector	['vektə]	n. 向量，矢量
		vt. 无线电导引
continuously	[kən'tinjuəsli]	adv. 不断地，连续地
robust	[rə'bʌst]	adj. 健壮的，精力充沛的
stimulus	['stimjuləs]	n. 刺激物，促进因素，刺激
decidedly	[di'saididli]	adv. 果断地，断然地
asymmetrical	[æsi'metrikəl]	adj. 不均匀的，不对称的

Phrases

a suite of	一套，一系列
wireless light switch	无线光交换机
electrical meter	电子仪表
be intended to be	规定为，确定为

high reliability	高可靠性
flash memory	闪存
build upon	指望，依赖，建立于
underlying structure	底层结构
powerline networking	电力线组网
smart metering	智能仪表
smart appliance	智能器具
device type	设备类型
pass on	传递
mains supply	干线供电，市电电源，交流电源
duty cycle	工作比
conform to	符合，遵照
access mode	存取方式，访问方式

Abbreviations

kbps (kilobits per second)	千位/秒
msec (millisecond)	毫秒
CSMA (Carrier Sense Multiple Access)	载波侦听多路访问
GTS (Guaranteed Time Slot)	保障时隙

Exercises

I. Answer the following questions according to the text.

1. What is ZigBee?
2. What are the Industrial, Scientific and Medical (ISM) radio bands ZigBee operates in?
3. What must the coordinator be within star networks?
4. Why can the latency be low and devices can be responsive?
5. Why can average power consumption be low, resulting in long battery life?
6. How many types of ZigBee devices are mentioned in the passage? What are they?
7. What do the current ZigBee protocols support?
8. What is used in non-beacon-enabled networks?
9. What are the special network nodes called in beacon-enabled networks? What do they do?
10. What are the three notable exceptions to the use of CSMA?

II. Translate the following terms or phrases from English into Chinese and vice versa.

1. Wireless Embedded Internet   1. _____
2. beehive   2. _____
3. 高可靠性   3. _____

4. functionality   4. _____
5. vector   5. _____
6. 闪存   6. _____
7. smart metering   7. _____
8. wireless light switch 8. _____
9. periodic   9. _____
10. latency   10. _____

Text A 参考译文

IEEE 802.15.4

IEEE 802.15.4 是一个标准，它为低速率无线个人局域网(LR-WPANs)中物理层和介质访问控制提供规范，它由 IEEE 802.15 工作组维护。它是 ZigBee、ISA100.11a、Wireless HART 以及 MiWi 规范的基础，这些规范都进一步扩展了该标准，方法是开发 802.15.4 没有定义的上层。另外一个选择是将 IEEE 802.15.4 和 6LoWPAN 以及标准的因特网协议一起使用，建立一个无线嵌入式因特网(如图 9.1 所示，图略)。

1. 概论

IEEE 802.15.4 标准打算为无线个人局域网(WPAN)提供基础较低的网络层。WPAN 着重在低成本、低速设备之间广泛通信(与之相对的是其他更面向终端用户的方法，如 WiFi)，重点是近距离的、有很少或无底层基础设施的设备之间的低成本通信，且打算用它来降低功耗。

基本框架设想在 10 m 的通信范围内，传输率为 250 kb/s，也可以折衷传输率来支持更多功耗更低的嵌入设备，方法是定义多个而不是一个物理层。初期定义的传输率更低，在 20 kb/s 到 40 kb/s 之间，最近的修订版中的传输率已经增加了 100 kb/s。

使用低传输率主要是为了降低功耗。如上所述，WPAN 中 802.15.4 的主要标志性特点在于制造和运行成本低，而且技术简单，不会牺牲灵活性和通用性。

实时适应性也是 IEEE 802.15.4 的主要特色，方法是保留确定的时隙、通过 CSMS/CA 避免冲突并完全支持安全通信。802.15.4 设备也包括电源管理功能，如链接质量和电能检查。

802.15.4 兼容设备使用三种频段中的一种来运行。

2. 协议体系

802.15.4 设备可以通过概念简单的无线网络实现相互结合。网络层的定义基于 OSI 模型；虽然在该标准中只定义了低层，但如果打算与高层交互，则可使用 IEEE 802.2 逻辑链路控制子层并通过汇聚子层来访问 MAC。此访问可依赖外部设备或者纯嵌入的、自运行设备来实现。

2.1 物理层

物理层(PHY)总的来说可提供数据传输服务，也可以作为物理层管理实体的接口，管理每层并维护相关个人局域网的信息库。因此，PHY 管理物理 RF 收发器并选择通道、管理能源和信号。它运行在下面三个可能未经许可的波段之一：

- 868.0～868.6 MHz：欧洲，允许一个信道(2003, 2006)；
- 902～928 MHz：北美，高达 10 个通道(2003)，已扩展到 30 个(2006)；
- 2400～2483.5 MHz：全球使用，高达 60 通道(2003, 2006)。

该标准的 2003 年版本原来指定了基于直接序列扩频(DSSS)技术的两个物理层：一个工作在传输率为 20～40 kb/s 的 868/915 MHz 波段，另一个工作在传输率为 250 kb/s 的 2450 MHz 波段。

2006 年版本增大了 868/915 MHz 波段的数据传输率，上调到支持 100 kb/s 和 250 kb/s 的传输率。更重要的是，它进一步定义了基于所使用的调制模式的四个物理层。其中三个层继续使用 DSSS 方法：在 868/915 MHz 波段，使用二相或偏移四相移相键控(其中第二个是一个备选项)；在 2450 MHz 波段，使用偏移四项移相键控。另外的选择是，用二相键控和幅移键控组合定义 868/915 MHz 层(因此这就基于平行，而不是序列扩展频谱，PSSS)。在支持的 868/915 MHz PHY 之间动态切换是可能的。

除了这三个波段外，IEEE 802.15.4c 研究组正在考虑使用最新在中国开放的 314～316 MHz、430～434 MHz 以及 779～787 MHz 波段，而 IEEE 802.15 Task Group 4d 正在改进现有的 802.15.4-2006 标准以便支持日本新的 950～956 MHz 波段。这些组提出的第一版修订标准已经于 2009 年 4 月发布。

在 2007 年 8 月, IEEE 802.15.4a 已经发布，它把早期 2006 版的四个 PHY 扩大到六个，增加了一个使用直接序列超宽带的 PHY(UWB)和另一个使用线性调频技术的 PHY(CSS)。给 UWB PHY 分配的三个频率范围为：低于 1 GHz、3～5 GHz 以及 6～10 GHz 的频率。给 CSS PHY 分配的频谱为 2450 MHz 波段。

在 2009 年 4 月，发布了 IEEE 802.15.4c 和 IEEE 802.15.4d，它们扩展了另外几个可用的 PHY：一个使用 O-QPSK 或 MPSK 的 780 MHz 波段，另一个使用 GFSK 或 BPSK 的 950 MHz 波段。

2.2 MAC 层

介质访问控制能够通过物理通道传输 MAC 帧。除了数据服务外，MAC 层也会管理接口并且它本身也管理对物理通道和网络信标的访问。它也可以控制帧确认、保证时隙并处理节点连接。最后，它为安全服务提供钩入点。

2.3 更高层

在 IEEE 802.15.4 标准中没有定义其他更高的层和协作子层。目前已经有一些建立在这个标准上的规范(如 ZigBee)提出了整合解决方案。

3. 网络模型

3.1 节点类型

IEEE 802.15.4 标准定义了两类网络节点。

第一类是全功能设备(FFD)。它可以作为个人局域网的协调者,也起普通节点的作用。它实现了可以与其他任何设备交谈的普通通信模型;它也能转发消息,此时它被称为协调者。

另外一类是精简功能设备(RFD)。这就意味着它是资源不多、通信很少、功能简单的设备。因此,它们只能与FFD通信,绝对不能作为协调者。

3.2 拓扑

网络可以用点对点或星型拓扑结构组建(如图9.2所示,图略)。但是,每个网络都至少需要一个FFD起网络协调者的作用。网络就是这样由距离适当的一组组独立设备组成的。每个设备都有唯一的64位标识符。如果满足某些条件,则在受限制的环境中也可以使用16位标识符,也即,在每个PAN区域中,允许使用短标识符通信。

对等网(点对点网)可以组成专用的连接模式,其扩展只受每一对节点之间距离的限制。其目的是能够为自主管理与组织的特殊网络提供基础。因为IEEE 802.15.4标准没有定义网络层,所以不直接支持路由,但这样的一个附加层可以增加对多次反射通信的支持。以后也许会进一步增加拓扑限制。IEEE 802.15.4标准实际上使用的是簇树结构(如图9.3所示,图略),其中,一个RFD一次只能连接一个FFD来构成网络,只有RFD是树叶并且绝大多数节点是FFD。该结构可以扩展为普通掩膜网,其节点都是簇树网络,每个簇除了有全局协调者外,也有一个本地协调者。

该网络也支持更结构化的星型模式,但该网络的协调者必须是中心节点。在一个FFD选择唯一的PAN标识来建立自己的PAN并声明自己做协调者后,这种网络就出现了。此后,其他设备就可以连接到该网络了,这个网络完全独立于其他所有星型网络。

4. 数据传输体系

帧是数据传输的基本单位,它有四个基本类型(数据、确认、信标和MAC命令帧),在简明性和健壮性之间提供了合理的折衷。另外,也可以使用由协调者定义的超级帧结构,在这种情况下,两个信标作为界限并提供与其他设备同步的信息以及配置信息。超级帧由16个等长度的时隙组成,可以进一步划分为活动部分和非活动部分,在此期间协调者可以进入省电模式,无需控制其网络。

在超级帧中,发生在它们的界限之间的争用由CSMA/CA解决。每个传输必须在下一个信标到达之前结束。如上所述,明确带宽的应用要使用一个或多个无争用的时隙的7个域,这7个域跟在超级帧的末尾。超级帧的第一部分必须能够为网络结构及其设备提供服务。通常情况下,超级帧用于低延迟的设备,即使长期不活动,这些设备也必须保持联系。

给协调者传输数据需要一个信标同步期。如果合适的话,信标跟在CSMA/CA传输之后(如果使用超级帧,依靠多个时隙),也可跟在确认之后。来自协调者的数据传输通常遵从设备请求:如果使用信标,就用它们来发出请求信号;协调者确认该请求,然后以包的形式发送设备确认的数据。不用超级帧也一样,只是在这种情况下无需用信标来跟踪待审消息。

点对点网络既可以使用非时隙CSMA/CA,也可以使用同步机制。在这种情况下,任意两个设备之间都可以通信;而在"结构化"模式中,必须有一个设备是网络协调者。

通常，所有的执行过程都应遵循典型的请求-确认/指示-响应分类。

5. 可靠性与安全性

通过 CSMA/CA 协议可访问物理介质。不用信标机制的网络利用的是基于介质监听的无时隙变量，并使用随机指数后退算法；确认则不遵循这个准则。使用信标时，一般数据传输利用未分配的时隙，同样，确认也不遵循同样的进程。

在某种情况下，确认消息也是可选的。无论在何种情况下，如果一个设备不能在一定的时间内处理帧，就完全不能确定接收：基于超时的重发可能执行多次，然后决定放弃或者再试。

由于这些设备的预设环境假设电池寿命最大化，因此该协议力求处理所有帧，并定期检查待审消息，其频繁程度依应用的需要而定。

对于安全通信，MAC 子层提供了通过上层来驾驭的便利，以便实现期望的安全水平。更高层的处理也许要特定的钥匙来执行对称加密，以便保护有效荷载并限制它只可以用于一组设备或者只实现点对点链接。这些设备组可以在访问控制列表中指定。此外，在连续接收信息期间，MAC 计算不断进行检测以便确保老的帧或被认为无效的数据不转发到更高层。

除这种安全模式外，还有一种不安全的 MAC 模式，它通过访问控制列表并根据帧的(推测)来源决定是否接收帧。

Unit 10　Cloud Computing

How Cloud Computing Works?

1. Introduction

Let's say you're an executive at a large corporation. Your particular responsibilities include making sure that all of your employees have the right hardware and software they need to do their jobs. Buying computers for everyone isn't enough—you also have to purchase software or software licenses to give employees the tools they require. Whenever you have a new hire, you have to buy more software or make sure your current software license allows another user. It's so stressful that you find it difficult to go to sleep on your huge pile of money every night.

Soon, there may be an alternative for executives like you. Instead of installing a suite of software for each computer, you'd only have to load one application. That application would allow workers to log into a Web-based service which hosts all the programs the user would need for his or her job. Remote machines owned by another company would run everything from E-mail to word processing to complex data analysis programs. It's called cloud computing, and it could change the entire computer industry.

In a cloud computing system(See Figure 10.1), there's a significant workload shift. Local computers no longer have to do all the heavy lifting when it comes to running applications. The network of computers that make up the cloud handles them instead. Hardware and software demands on the user's side decrease. The only thing the user's computer needs to be able to run is the cloud computing system's interface software, which can be as simple as a Web browser, and the cloud's network takes care of the rest.

There's a good chance you've already used some form of cloud computing. If you have an e-mail account with a Web-based E-mail service like Hotmail, Yahoo! Mail or Gmail, then you've had some experience with cloud computing. Instead of running an E-mail program on your computer, you log in to a Web E-mail account remotely. The software and storage for your account doesn't exist on your computer—it's on the service's computer cloud.

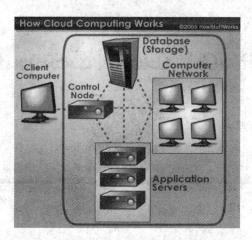

Figure 10.1　A typical cloud computing system

2. Cloud Computing Architecture

When talking about a cloud computing system, it's helpful to divide it into two sections: the front end and the back end. They connect to each other through a network, usually the Internet. The front end is the side the computer user, or client, sees. The back end is the "cloud" section of the system.

The front end includes the client's computer (or computer network) and the application required to access the cloud computing system. Not all cloud computing systems have the same user interface. Services like Web-based E-mail programs leverage existing Web browsers like Internet Explorer or Firefox. Other systems have unique applications that provide network access to clients.

On the back end of the system are the various computers, servers and data storage systems that create the "cloud" of computing services. In theory, a cloud computing system could include practically any computer program you can imagine, from data processing to video games. Usually, each application will have its own dedicated server[①].

A central server administers the system, monitoring traffic and client demands to ensure everything runs smoothly. It follows a set of rules called protocols and uses a special kind of software called middleware[②]. Middleware allows networked computers to communicate with each other. Most of the time, servers don't run at full capacity. That means there's unused processing power going to waste. It's possible to fool a physical server into thinking it's actually multiple servers, each running with its own independent operating system. The technique is

---

① In most common use, server is a physical computer (a computer hardware system) dedicated to running one or more such services (as a host), to serve the needs of users of the other computers on the network. Depending on the computing service that it offers it could be a database server, file server, mail server, print server, web server, or some other kind of server.

② In the computer industry, middleware is a general term for any programming that serves to "glue together (胶合)" or mediate between two separate and often already existing programs.

called server virtualization. By maximizing the output of individual servers, server virtualization reduces the need for more physical machines.

If a cloud computing company has a lot of clients, there's likely to be a high demand for a lot of storage space. Some companies require hundreds of digital storage devices. Cloud computing systems need at least twice the number of storage devices it requires to keep all its clients' information stored. That's because these devices, like all computers, occasionally break down. A cloud computing system must make a copy of all its clients' information and store it on other devices. The copies enable the central server to access backup machines to retrieve data that otherwise would be unreachable. Making copies of data as a backup is called redundancy.

3. Cloud Computing Applications

The applications of cloud computing are practically limitless. With the right middleware, a cloud computing system could execute all the programs a normal computer could run. Potentially, everything from generic word processing software to customized computer programs designed for a specific company could work on a cloud computing system.

Why would anyone want to rely on another computer system to run programs and store data? Here are just a few reasons:

• Clients would be able to access their applications and data from anywhere at any time. They could access the cloud computing system using any computer linked to the Internet. Data wouldn't be confined to a hard drive on one user's computer or even a corporation's internal network.

• It could bring hardware costs down. Cloud computing systems would reduce the need for advanced hardware on the client side. You wouldn't need to buy the fastest computer with the most memory, because the cloud system would take care of those needs for you. Instead, you could buy an inexpensive computer terminal. The terminal could include a monitor, input devices like a keyboard and mouse and just enough processing power to run the middleware necessary to connect to the cloud system. You wouldn't need a large hard drive because you'd store all your information on a remote computer.

• Corporations that rely on computers have to make sure they have the right software in place to achieve goals. Cloud computing systems give these organizations company-wide access to computer applications. The companies don't have to buy a set of software or software licenses for every employee. Instead, the company could pay a metered fee to a cloud computing company.

• Servers and digital storage devices take up space. Some companies rent physical space to store servers and databases because they don't have it available on site. Cloud computing gives these companies the option of storing data on someone else's hardware, removing the need for physical space on the front end.

• Corporations might save money on IT support. Streamlined hardware would, in theory, have fewer problems than a network of heterogeneous machines and operating systems[①].

• If the cloud computing system's back end is a grid computing system[②], then the client could take advantage of the entire network's processing power. Often, scientists and researchers work with calculations so complex that it would take years for individual computers to complete them. On a grid computing system, the client could send the calculation to the cloud for processing. The cloud system would tap into the processing power of all available computers on the back end, significantly speeding up the calculation.

4. Cloud Computing Concerns

Perhaps the biggest concerns about cloud computing are security and privacy. The idea of handing over important data to another company worries some people. Corporate executives might hesitate to take advantage of a cloud computing system because they can't keep their company's information under lock and key.

The counterargument to this position is that the companies offering cloud computing services live and die by their reputations. It benefits these companies to have reliable security measures in place. Otherwise, the service would lose all its clients. It's in their interest to employ the most advanced techniques to protect their clients' data.

Privacy is another matter. If a client can log in from any location to access data and applications, it's possible the client's privacy could be compromised. Cloud computing companies will need to find ways to protect client privacy. One way is to use authentication[③] techniques such as user names and passwords. Another is to employ an authorization format—each user can access only the data and applications relevant to his or her job.

---

① An operating system is the most important program that runs on a computer. Every general-purpose computer must have an operating system to run other programs. Operating systems perform basic tasks, such as recognizing input from the keyboard, sending output to the display screen, keeping track of files and directories ([di'rektəri] n. 目录) on the disk, and controlling peripheral devices such as disk drives and printers.

② Grid computing is a term referring to the federation of computer resources from multiple administrative domains to reach a common goal. The grid can be thought of as a distributed system with noninteractive (['nʌnintə'æktiv] adj.非交互的) workloads that involve a large number of files. What distinguishes grid computing from conventional high performance computing systems such as cluster computing is that grids tend to be more loosely coupled, heterogeneous, and geographically dispersed ([dis'pə:st] adj. 被分散的). Although a grid can be dedicated to (专注于) a specialized application, it is more common that a single grid will be used for a variety of different purposes. Grids are often constructed with the aid of general-purpose grid software libraries known as middleware.

③ The process of identifying an individual, usually based on a username and password. In security systems, authentication is distinct ([dis'tiŋkt] adj. 截然不同的,) from authorization, which is the process of giving individuals access to system objects based on their identity ([ai'dentiti] n. 身份). Authentication merely ensures that the individual is who he or she claims to be, but says nothing about the access rights of the individual.

Some questions regarding cloud computing are more philosophical. Does the user or company subscribing to the cloud computing service own the data? Does the cloud computing system, which provides the actual storage space, own it? Is it possible for a cloud computing company to deny a client access to that client's data? Several companies, law firms and universities are debating these and other questions about the nature of cloud computing.

New Words

executive	[ig'zækjutiv]	n. 总经理；行政部门；[计算机]执行指令
		adj. 执行的；管理的；政府部门的
corporation	[kɔːpə'reiʃən]	n. 公司；企业
responsibility	[ris,pɔnsə'biliti]	n. 责任，职责
employee	[,emplɔ'iː]	n. 雇工，雇员，职工
license	['laisəns]	n. 许可(证)，执照
		v. 许可
hire	['haiə]	n. 租金；酬金，工钱；[非正式用语] 被雇佣的人
		vt. 聘用；录用；雇用；租用
		vi. 受雇；得到工作
stressful	['stresful]	adj. 产生压力的，使紧迫的
remote	[ri'məut]	adj. 远程的，遥远的
workload	['wəːkləud]	n. 工作量
handle	['hændl]	vt. 处理，操作
		n. 句柄
interface	['intə,feis]	n. 接口，界面
browser	[brauzə]	n. 浏览器
account	[ə'kaunt]	n. 账户
experience	[iks'piəriəns]	n. & vt. 经验，体验，经历
storage	['stɔridʒ]	n. 存储
leverage	['liːvəridʒ]	n. 手段；杠杆作用
server	['səːvə]	n. 服务器
middleware	['midlwɛə]	n. 中间设备，中间件
occasionally	[ə'keiʒənəli]	adv. 有时候，偶尔
unreachable	[ʌn'riːtʃəbl]	adj. 取不到的，不能得到的
redundancy	[ri'dʌndənsi]	n. 冗余
limitless	['limitlis]	adj. 无限的，无界限的
potentially	[pə'tenʃəli]	adv. 潜在地
digital	['didʒitl]	adj. 数字的，数位的
		n. 数字，数字式

remove	[ri'mu:v]	vt. 消除，删除，去除
streamlined	['stri:mlaind]	adj. 最新型的，改进的
heterogeneous	[ˌhetərəu'dʒi:niəs]	adj. 不同种类的，异类的
worry	['wʌri]	n. 烦恼，忧虑
		vt. 使烦恼，使焦虑；困扰，折磨
hesitate	['heziteit]	v. 犹豫，踌躇，不愿
counterargument	['kauntəˌɑ:gjə:mənt]	n. 辩论，抗辩
philosophical	[filə'sɔfikəl]	adj. 哲学的；达观的
reputation	[ˌrepju(:)'teiʃən]	n. 名誉，名声
compromise	['kɔmprəmaiz]	v. 危及……的安全
authentication	[ɔːˌθenti'keiʃən]	n. 认证，证明，鉴定
authorization	[ˌɔ:θərai'zeiʃən]	n. 授权，认可
deny	[di'nai]	v. 否认，拒绝

Phrases

cloud computing	云计算
make sure	确保，确定；确信，证实
word processing	字处理
data analysis program	数据分析程序
computer industry	计算机行业
interface software	接口软件
on the side	另外
take care of	承担
front end	前端
back end	后端
data processing	数据处理
video game	视频游戏
at full capacity	满功率，满负载
fool sb. into doing sth.	哄骗某人做某事
go to waste	浪费掉，白费
server virtualization	服务器虚拟化
storage space	存储空间
digital storage device	数字化存储设备
break down	毁掉，停顿，中止，垮掉
take up	占用
be confined to	限制在，局限于
hard drive	硬盘驱动器
grid computing system	网格计算系统

tap into	挖掘
under lock and key	严密保管，妥善保管，妥善存放
law firm	律师事务所

Exercises

I. Answer the following questions according to the text.

1. What do your particular responsibilities include as an executive at a large corporation?
2. How many sections can a cloud computing system? What are they?
3. What does the front end include?
4. What does the back end include?
5. How does server virtualization reduce the need for more physical machines?
6. Why do Cloud computing systems need at least twice the number of storage devices it requires to keep all its clients' information stored?
7. What is called redundancy?
8. Would you need to buy the fastest computer with the most memory? Why?
9. Why do some companies rent physical space to store servers and databases?
10. What are the biggest concerns about cloud computing? How can Cloud computing companies protect client privacy?

II. Translate the following terms or phrases from English into Chinese and vice versa.

1. cloud computing　　　　　　　　1. _____
2. n. 浏览器　　　　　　　　　　　2. _____
3. n. 服务器　　　　　　　　　　　3. _____
4. remote　　　　　　　　　　　　4. _____
5. grid computing system　　　　　 5. _____
6. 数字化存储设备　　　　　　　　6. _____
7. interface　　　　　　　　　　　7. _____
8. authentication　　　　　　　　　8. _____
9. storage　　　　　　　　　　　　9. _____
10. n. 句柄　　　　　　　　　　　 10. _____

III. Fill in the blanks with the words given below.

survives	maintenance	manipulate	supply	warehouse
retrieve	space	messages	same	redundancy

❖❖❖ How Cloud Storage Works ❖❖❖

There are hundreds of different cloud storage systems. Some have a very specific focus, such as storing Web E-mail ___1___ or digital pictures. Others are available to store all forms of digital data. Some cloud storage systems are small operations, while others are so large that the physical equipment can fill up an entire ___2___. The facilities that house cloud storage systems are called data centers.

At its most basic level, a cloud storage system needs just one data server connected to the Internet. A client (e.g., a computer user subscribing to a cloud storage service) sends copies of files over the Internet to the data server, which then records the information. When the client wishes to ___3___ the information, he or she accesses the data server through a Web-based interface. The server then either sends the files back to the client or allows the client to access and ___4___ the files on the server itself.

Cloud storage systems generally rely on hundreds of data servers. Because computers occasionally require ___5___ or repair, it's important to store the same information on multiple machines. This is called redundancy. Without ___6___, a cloud storage system couldn't ensure clients that they could access their information at any given time. Most systems store the ___7___ data on servers that use different power supplies. That way, clients can access their data even if one power ___8___ fails.

Not all cloud storage clients are worried about running out of storage ___9___. They use cloud storage as a way to create backups of data. If something happens to the client's computer system, the data ___10___ off-site. It's a digital-age variation of "don't put all your eggs in one basket".

IV. Translate the following passage from English to Chinese.

For some computer owners, finding enough storage space to hold all the data they've acquired is a real challenge. Some people invest in larger hard drives. Others prefer external storage devices like thumb drives or compact discs. Desperate computer owners might delete entire folders worth of old files in order to make space for new information. But some are choosing to rely on a growing trend: cloud storage.

While cloud storage sounds like it has something to do with weather fronts and storm systems, it really refers to saving data to an off-site storage system maintained by a third party. Instead of storing information to your computer's hard drive or other local storage device, you save it to a remote database. The Internet provides the connection between your computer and the database.

On the surface, cloud storage has several advantages over traditional data storage. For example, if you store your data on a cloud storage system, you'll be able to get to that data from any location that has Internet access. You wouldn't need to carry around a physical storage device or use the same computer to save and retrieve your information. With the right storage system, you could even allow other people to access the data, turning a personal project into a collaborative effort.

4G—The Future of Mobile Internet

1. How fast is 4G?

1.1 What Does 4G Mean to You?

4G simply means fourth generation in mobile technologies. It is the common name for IMT (International Mobile Telecommunications) Advanced standards defined by International Telecommunications Union (ITU). What does the standard mean to common man? It means an awesome one Gigabit per second of download speed, while you access the internet, through wireless means! When both the transmitting and receiving station are mobile, the speed is a bit less. When mobile, that is at a speed of at least 60 Mph, 4G offers you a download speed of 100 Megabits per second. More bluntly, 4G allows you to download an HD movie in no more than 30 seconds! ITU-Radio communications defines IMT-Advanced (4G) as a network which provides a global platform on which to build the next generations of mobile services—fast data access, unified messaging and broadband multimedia—in the form of exciting new interactive services.

1.2 The Fourth Generation of Mobile Technologes

Now more pertinent questions, there are a lot of technologies offering high speed internet, how would one know if he/she is using this 4G stuff or not? Well, ITU has specified certain technical standards, which should be met by any technology that claims to be 4G. The specifications list the minimum data transfer rate that the proposed 4G network should have. The 4G working group has defined the following as objectives of the 4G wireless communication standard:

• A nominal data rate of 100 Mb/s while the client physically moves at high speeds relative to the station, and 1 Gb/s while client and station are in relatively fixed positions as defined by the ITU-R.

• A data rate of at least 100 Mb/s between any two points in the world.

• Flexible channel bandwidth, between 5 and 20 MHz, optionally up to 40 MHz.

• Peak link spectral efficiency of 15 b/s/Hz in the downlink, and 6.75 b/s/Hz in the uplink (meaning that 1 Gb/s in the downlink should be possible over less than 67 MHz bandwidth).

• System spectral efficiency of up to 3 b/s/Hz/cell in the downlink and 2.25 b/s/Hz/cell for indoor usage.

• High quality of service for next generation multimedia support (real time audio, high speed data, HDTV video content, mobile TV, etc.).

• Interoperability with existing wireless standards, and an all IP, packet switched network.

• Smooth hand-off, seamless connectivity and global roaming across multiple networks.

2. What is the Importance of 4G?

What makes 4G, or for that matter wireless broadband, inevitable? How is 4G going to bring changes in everything related to internet? The last specification mentioned in the ITU-R standard makes it more attractive: the data transfer through the network is using packets, using IP or internet protocol. The revolution brought out by Web 2.0① is quite conspicuous to be missed. User generated content literally ruled the web; portals and social networking sites② like youtube and facebook became nothing less than cultural icons. The web is still improving. Interactive applications like IPTV just changed our concept of television broadcasting. All these are going to radically improve, perhaps, beyond our imagination, once we switch over to 4G networks. The major challenge these applications are facing is not the technology of themselves, but that of networks-information bottlenecks. Upload speeds are still a concern, all the more if it is a mobile platform. Here is where 4G comes into play with its amazing one gigabit per second data rate. The ultra-fast 4G could truly revolutionize our concept of interactive applications.

3. The Future of 4G and You

Cloud computing is another area, which is going to really cash on switching over to 4G networks. Increased speed, mobility and portability that 4G offers, is a shot in the arm for cloud computing applications. When people avoid capital expenditure on resources, the increased data transfer rate will be a must for applications to work smoothly. If you are saving your data on a server or running your favorite application from the server (because you do not want a hard disk carried with you, how much time can you afford to get the data down? Even couple of seconds would make the cloud computing unattractive. Cloud computing thus enables mobile phone sized stations to be acted as powerful computers, if and only if they are connected to something that offers a 4G speed in upload as well as download. The mobility offered makes the whole thing personal and portable, all the while being cost effective. Imagine, using the whole operating system itself through the cloud computing architecture. 4G makes this absolutely possible!

Internet of Things (IoT) is the next big thing to happen. It is dubbed as the future in which physical objects, even in our everyday life, are going to get networked. It is something similar to the present day RFID cards and other types of identification technologies. The day will not be far

---

① Web 2.0 is a loosely defined intersection of web application features that facilitate participatory information sharing (信息共享), interoperability, user-centered design, and collaboration on the World Wide Web. A Web 2.0 site allows users to interact and collaborate with each other in a social media dialogue as creators of user-generated content (用户自创内容) in a virtual community (虚拟社区), in contrast to websites where users are limited to the passive viewing of content that was created for them. Examples of Web 2.0 include social networking sites, blogs ([blɔg] n. 博客), wikis, video sharing sites, hosted services, web applications, mashups and folksonomies.

② A Web site that provides a virtual community for people to share their daily activities with family and friends, or to share their interest in a particular topic, or to increase their circle of acquaintances ([ə'kweintəns]n. 相识，熟人). There are dating sites, friendship sites, sites with a business purpose and hybrids (['haibrid] n. 混合，adj.混合的) that offer a combination of these.

when our pen, book, shirts, shoes or parts of automobiles are equipped with minuscule identifying chips which exchanges information between them. Imagine the flow of information. It can doubtlessly be said 4G will be a significant step towards such an internet of things.

4. How 4G can change your life?

National broadband plan of the United States Federal Communications Commission envisions how a paradigmatic shift can be brought about in a number of areas like healthcare, education, economic opportunities, energy and environment, government performance, civic engagement and public safety by adopting broadband. Markets will be given new opportunities and consumers can avail new products. Using 4G, classroom experience can be dramatically altered. Both inside out and outside classrooms, learning can be more fun and interactive. Interactive applications and faster knowledge sharing will make healthcare cost-effective and time-saving. Modernization of electric grid would mean more energy conservation. 4G is environment friendly when the transmission power is utilized to carry more data than previous generation technologies practically carried. The world is moving towards e-governance①. Better service delivery through 4G technology is something that is desired across the globe. Broadband services enable the citizen to effectively utilize his right to information concerning government information. Another important feature of the 4G standards will be improved security of data. Due to its emphasis on robustness and security, the concerns about data theft would be minimized.

In summary, 4G can be said to give the best that a wireless network is desired of. It gives ultra hi-speed internet with portability, reliability and easy accessibility. While download and upload limits and speed reductions continue to worry our daily routines, adopting 4G is inevitable as well as imminent.

New Words

awesome	['ɔːsəm]	adj. 出色的，非凡的，可怕的
download	['daunləud]	v. 下载
wireless	['waiəlis]	adj. 无线的
bluntly	['blʌntli]	adv. 坦率地，率直地
platform	['plætfɔːm]	n. 平台
interactive	[ˌintər'æktiv]	adj. 交互式的
pertinent	['pəːtinənt]	adj. 有关的，相干的，中肯的

---

① E-Governance is the application of Information and Communication Technology (ICT) for delivering government services, exchange of information communication transactions, integration various stand-one systems and services between <u>Government-to-Citizens</u> (G2C，政府对市民), <u>Government-to-Business</u> (G2B，政府对企业), <u>Government-to-Government</u> (G2G，政府对政府) as well as back office processes and interactions within the entire government frame work.

propose	[prəˈpəuz]	vt. 计划，建议，向……提议
station	[ˈsteiʃən]	n. 基站
spectral	[ˈspektrəl]	adj. 光谱的，频谱的
downlink	[ˈdaunliŋk]	n. 下行线
seamless	[ˈsiːmlis]	adj. 无缝的
connectivity	[kənekˈtiviti]	n. 连通性
importance	[imˈpɔːtəns]	n. 重要(性)，重大，价值
broadband	[ˈbrɔːdbænd]	n. 宽带
inevitable	[inˈevitəbl]	adj. 不可避免的，必然的
attractive	[əˈtræktiv]	adj. 吸引人的，有魅力的
conspicuous	[kənˈspikjuəs]	adj. 显著的
portal	[ˈpɔːtəl]	n. 门户，入口
bottleneck	[ˈbɔtl,nek]	n. 瓶颈
upload	[ˈʌp,ləud]	v. 上传，上载
gigabit	[ˈdʒigəbit]	n. 吉比特
mobility	[məuˈbiliti]	n. 活动性，灵活性，机动性
unattractive	[ˌʌnəˈtræktiv]	adj. 不吸引人的
doubtless	[ˈdautlis]	adj. 无疑的，确定的
		adv. 无疑地，确定地
envision	[inˈviʒən]	vt. 想象，预想
paradigmatic	[ˌpærədigˈmætik]	adj. 范式的，例证的
avail	[əˈveil]	vi. 有益于，有帮助，有用，有利
		vt. 有利于
		n. 效用，利益
dramatically	[drəˈmætikəli]	adv. 戏剧地，引人注目地
e-governance	[iː-ˈgʌvənəns]	n. 电子治理，电子管理，电子政务
minimize	[ˈminimaiz]	vt. 将……减到最少
		v. 最小化
accessibility	[ˌækəsesiˈbiliti]	n. 易访问的，可到达的
routine	[ruːˈtiːn]	n. 日常事务，例行公事，常规，程序
imminent	[ˈiminənt]	adj. 即将来临的，逼近的

Phrases

mobile Internet	移动因特网
common man	平民
Gigabit per second	每秒吉字节
data transfer rate	数据传输率
nominal data rate	标称数据率

channel bandwidth	频道带宽，通道带宽
spectral efficiency	频谱效率，频谱利用率
global roaming	国际漫游
social networking site	社交网站
cultural icon	文化符号，文化图腾，文化标志
be going to	要，会，将要
switch over	转变
capital expenditure	资本支出，基本建设费用
be equipped with	装备
United States Federal Communications Commission	美国联邦通信委员会
paradigmatic shift	范式变革
energy conservation	能源节约
put emphasis on	强调
data theft	数据失窃

Abbreviations

4G (Fourth Generation)	第四代移动通信及其技术
IMT (International Mobile Telecommunications)	国际移动通信
ITU (International Telecommunications Union)	国际电信联盟
Mph (Megabits per second)	每秒兆字节
HD (High Definition)	高清晰度
HDTV (High Definition TV)	高清晰度电视
ITU-R (Radio communication Sector of ITU)	国际电信联盟无线电通信部门

Exercises

I. Fill in the blanks with the information given in the text.

1. 4G simply means _____. It is the common name for IMT (International Mobile Telecommunications) Advanced standards defined by _____.

2. 4G allows you to download an HD movie in _____.

3. The minimum data transfer rate that the proposed 4G network should have listed by the specifications is a data rate of _____ between any two points in the world.

4. Portals and social networking sites like _____ and _____ became nothing less than cultural icons.

5. The major challenge these applications are facing is not _____, but _____.

6. When people avoid _____, the increased data transfer rate will be a must for applications to _____.

7. Cloud computing thus enables mobile phone sized stations to be acted as _____, if and only if they are connected to something that offers a 4G speed in _____ as well as _____.

8. _____ is the next big thing to happen. It is dubbed as the future in which physical objects, even in our everyday life, are going to _____.

9. Using 4G, classroom experience can be _____. Both inside out and outside classrooms, learning can be _____.

10. Broadband services enable _____ to effectively utilize his right to information concerning _____.

II. Translate the following terms or phrases from English into Chinese and vice versa.

1. nominal data rate   1. _____
2. station   2. _____
3. data transfer rate   3. _____
4. adj. 无缝的   4. _____
5. channel bandwidth   5. _____
6. adj. 交互式的   6. _____
7. data theft   7. _____
8. social networking site   8. _____
9. n. 宽带   9. _____
10. downlink   10. _____

Text A 参考译文

云计算是如何工作的？

1. 绪论

假定你是大公司的主管，你负有特殊的责任，包括让全体员工都有适当的硬件和软件来满足其工作要求。但只为每人购买计算机是不够的——还要购买软件或软件许可，为员工提供所需要的工具。只要雇佣新员工，就必须购买更多的软件或确保现有的软件许可供另一个员工使用，以确保软件够用。压力如此之大，以至于你每夜都无法在"钱墩"上入睡。

对于像你这样的主管，不久就会有其他的选择了。你无需为每个计算机安装整套的软件，而是只装入一个应用程序。该应用程序让工作人员登录基于网络的服务，它为用户提供工作所需的全部程序。其他公司的远程机器可以运行各种程序，从电子邮件、文字处理到复杂的数据分析。这种服务叫云计算，它可能会改变整个计算机行业。

在云计算系统中(如图 10.1 所示，图略)，有一个重要的工作量转移。当运行应用程序时，本地计算机不用再承担全部繁重的任务，而是由构成云的云计算机网络去处理它们，所以用户端所需的软件和硬件会减少。用户的计算机只要能够运行云计算系统的接口软件(可以只是一个网络浏览器)即可，而云网络完成其余任务。

你可能已经使用了某种形式的云计算。如果你有一个基于网络电子邮件服务的电子邮件账号，如 Hotmail、Yahoo! Mail 或 Gmail 账号，那么你已经有了一些云计算体验。因为你不是在你的计算机上运行电子邮件程序，而是远程登录网络电子邮件账号。这些软件和你的账号并不存储在你的计算机上——而是存储在提供服务的计算机云中。

2. 云计算体系结构

当谈论云计算系统时，将其分为以下两部分是有益的：前端和后端。它们通过网络(通常是因特网)彼此连接。前端是计算机用户这一方，或称客户端；后端是该系统的"云"区域。

前端包括客户的计算机(或计算机网络)以及访问云系统所需的应用程序。并非所有的云计算系统都使用相同的用户接口，可以利用像 Internet Explorer 或 Firefox 这样的现有浏览器实现如基于网络的电子邮件程序这类服务。其他系统有为客户提供网络访问的特殊应用程序。

这个系统的后端是各种计算机、服务器和数据存储系统，它们建立了提供计算服务的"云"。理论上，一个云计算系统实际上可以包含你能够想象的任何计算机程序，从数据处理到视频游戏，但通常，每个应用程序会有自己的专门服务器。

由一个中心服务器管理该系统，监管流量和用户需求，以便确保每个程序可以平稳地运行。它遵循叫做协议的一组规则，并使用叫做中间件的特殊软件。中间件使联网的计算机彼此通信。大部分时间，服务器没有满载运行，这就意味着一些没有使用的处理能力被浪费了，且可能让一个物理服务器误认为它实际上是多个服务器。每个服务器都运行独立的操作系统，这个技术叫做服务器虚拟化。服务器虚拟化通过使每个服务器的输出最大化，减少了对更多物理机器的需求。

如果一个云计算公司有许多客户，则很可能需要许多存储空间。有些公司需要数百个数字存储设备。云计算系统需要的存储设备数量至少是它存储全部客户信息所需数量的两倍，那是因为这些设备像所有计算机一样，有时会损坏。一个云计算系统必须备份其全部客户的信息并存储在其他设备上。这些备份使得中心服务器能够访问备份机来恢复其他计算机不能访问的数据。制作数据的副本作为备份叫做冗余。

3. 云计算应用

云计算的应用几乎是无限的。使用适当的中间件，云计算系统可以在一般计算机上运行所有程序。在未来，从一般字处理软件到为特定公司定制的计算机程序都可以在云计算系统上运行。

为什么人们要依赖其他的计算机系统来运行程序和存储数据？有以下几个原因：

• 客户可以随时随地访问它们的应用程序和数据。他们可以用任何连入因特网的计算机访问云计算系统。数据不需要局限于存储在用户计算机的硬盘中，乃至公司的内部网络上。

• 它可以带来硬件成本的下降。云计算系统会减少客户端高级硬件的需求。你不需要购买内存最大、速度最快的计算机，因为云系统会满足你的需求。你可以只购买一个便宜的计算机终端。该终端包括显示器、像键盘和鼠标这样的输入设备以及仅仅够运行连接到云系统的中间件所需的处理能力。因为所有信息都存储在远程计算机上，所以不需要大的硬盘。

• 依靠计算机的大型公司必须确保具有合适的、能够实现其目标的软件。云计算系统可以使这些组织实现全公司对计算机应用软件的访问。这些公司无需为每个员工购买整套软件或软件许可，而是向云计算公司按计量付费。

• 服务器和数字存储设备占用空间。因为自己的可用空间不足，一些公司租用空间来存储服务器和数据库。云计算给这些公司提供在其他人的硬件上存储数据的选择，而无需前端的物理空间。

• 公司可以节省 IT 支持资金。在理论上，新型硬件要比传统机器和操作系统网络的故障更少。

• 如果云计算系统的后端是栅格计算系统，那么客户可以利用整个网络的处理能力。常常，科学家和研究人员需要进行非常复杂的计算，甚至可能要在单个计算机上花费数年才能完成。在一个栅格计算系统上，客户可以把计算发送给云来处理。云系统将挖掘后端所有可用的处理能力，这就显著提高了计算速度。

4. 云计算的关注点

云计算最关注的或许就是安全和隐私了。把重要数据交给其他公司处理的想法或许会引起一些人的担心。公司主管可能会犹豫要不要利用云计算系统，因为他们不能保证自己公司的信息会被严密保管。

辩护方认为，安全和隐私关乎提供云计算服务公司的生死声誉。这些公司采用十分可靠的安全措施才会对它们自己有益，否则将失去全部客户。为了自己的利益，他们会雇佣最高技术水准的专家来保护他们客户的数据。

隐私是另一要素。如果一个客户可以从任何位置登录来访问数据和应用程序，就可能危及客户的隐私。云计算公司必须找到保护客户隐私的方法。一个方法是使用如用户名和密码这样的认证技术；另一个方法是采用授权格式——每个用户只能访问与其工作相关的数据和应用程序。

有些关于云计算的问题更具有哲学意味。注册到云计算服务的用户或公司是否拥有数据？提供实际存储空间的云计算公司拥有这些数据吗？云计算可以拒绝客户对自己数据的访问吗？一些公司、律师事务所和大学正在讨论这些问题以及其他关于云计算本质的问题。

Unit 11 Smart City

Text A

Smart City

A smart city is an urban development[①] vision to integrate multiple Information and Communication Technology (ICT) and Internet of Things (IoT) solutions in a secure fashion to manage a city's assets. The city's assets include, but are not limited to, local departments' information systems, schools, libraries, transportation systems, hospitals, power plants, water supply networks, waste management, law enforcement, and other community services. The goal of building a smart city is to improve the quality of life by using urban informatics and technology to improve the efficiency of services and meet residents' needs. ICT allows city officials to interact directly with the community and the city infrastructure and to monitor what is happening in the city, how the city is evolving, and how to enable a better quality of life. Through the use of sensors integrated with real-time monitoring systems, data are collected[②] from citizens and devices and then processed and analyzed. The information and knowledge gathered are keys to tackling inefficiency.

ICT is used to enhance quality, performance and interactivity of urban services, to reduce costs and resource consumption and to improve contact between citizens and government. Smart city applications are developed with the goal of improving the management of urban flows and allowing for real time responses to challenges. A smart city may, therefore, be more prepared to respond to challenges than one with a simple "transactional" relationship with its citizens. Yet, the term itself remains unclear to its specifics and, therefore, open to many interpretations.

---

① Urban planning is a technical and political process concerned with the development and use of land, protection and use of the environment, public welfare (['welfɛə]n.福利), and the design of the urban environment, including air, water, and the infrastructure passing into and out of urban areas, such as transportation, communications, and distribution networks.
② Data collection is the process of gathering and measuring information on targeted variables in an established systematic fashion, which then enables one to answer relevant questions and evaluate outcomes (['autkʌm]n. 结果，成果).

Other terms that have been used for similar concepts include cyberville, digital city, electronic communities, flexicity, information city, intelligent city, knowledge-based city, and ubiquitous city.

Sectors that have been developing smart city technology include government services, transport and traffic management, energy, health care, water, innovative urban agriculture and waste management.

Major technological, economic and environmental changes have generated interest in smart cities, including climate change①, economic restructuring, the move to online retail and entertainment, urban population growth and pressures on public finances. The European Union (EU) has devoted constant efforts to devising a strategy for achieving "smart" urban growth for its metropolitan city-regions. The EU has developed a range of programmes under "Europe's Digital Agenda". In 2010, it highlighted its focus on strengthening innovation and investment in ICT services for the purpose of improving public services and quality of life. Arup estimates that the global market for smart urban services will be $400 billion per annum by 2020.

1. Terminology

Due to the breadth of technologies that have been implemented under the smart city label, it is difficult to distil a precise definition of a smart city. Deakin and Al Wear list four factors that contribute to the definition of a smart city:

- The application of a wide range of electronic and digital technologies to communities and cities;
- The use of ICT to transform life and working environments within the region;
- The embedding of such ICTs in government systems;
- The territorialisation of practices that brings ICTs and people together to enhance the innovation and knowledge that they offer.

Deakin defines the smart city as one that utilises ICT to meet the demands of the market (the citizens of the city), and that community involvement in the process is necessary for a smart city. A smart city would thus be a city that not only possesses ICT technology in particular areas, but also has implemented this technology in a manner that positively impacts the local community.

2. Characteristics

It has been suggested that a smart city (also community, business cluster②, urban agglomeration or region) uses information technologies to:

---

① Climate change is a change in the statistical distribution of weather patterns when that change lasts for an extended period of time (i.e., decades to millions of years). Climate change may refer to a change in average weather conditions, or in the time variation of weather around longer-term average conditions (i.e., more or fewer extreme weather events).

② A business cluster is a geographic concentration ([ˌkɔnsen'treiʃən]n.集中, 集合) of interconnected businesses, suppliers, and associated institutions in a particular field. Clusters are considered to increase the productivity with which companies can compete, nationally and globally.

(1) Make more efficient use of physical infrastructure (roads, built environment and other physical assets) through artificial intelligence and data analytics to support a strong and healthy economic, social, cultural development.

(2) Engage effectively with local people in local governance and decision by use of open innovation processes and e-participation, improving the collective intelligence of the city's institutions through e-governance, with emphasis placed on citizen participation and co-design.

(3) Learn, adapt and innovate and thereby respond more effectively and promptly to changing circumstances by improving the intelligence of the city.

They evolve towards a strong integration of all dimensions of human intelligence, collective intelligence, and also artificial intelligence within the city. The intelligence of cities "resides in the increasingly effective combination of digital telecommunication networks (the nerves), ubiquitously embedded intelligence (the brains), sensors and tags (the sensory organs), and software (the knowledge and cognitive competence)".

These forms of intelligence in smart cities have been demonstrated in three ways:

(1) Orchestration intelligence: Where cities establish institutions and community-based problem solving and collaborations, such as in Bletchley Park, where the Nazi Enigma cypher was decoded by a team led by Alan Turing①. This has been referred to as the first example of a smart city or an intelligent community.

(2) Empowerment intelligence: Cities provide open platforms, experimental facilities and smart city infrastructure in order to cluster innovation in certain districts. These are seen in the Kista Science City in Stockholm and the Cyberport Zone in Hong Kong. Similar facilities have also been established in Melbourne.

(3) Instrumentation intelligence: Where city infrastructure is made smart through real-time data② collection, with analysis and predictive modelling③ across city districts. There is much controversy surrounding this, particularly with regards to surveillance issues in smart cities. Examples of Instrumentation intelligence have been implemented in Amsterdam. This is implemented through:

- A common IP infrastructure that is open to researchers to develop applications;
- Wireless meters and devices transmit information at the point in time;

---

① Alan Mathison Turing (23 June 1912–7 June 1954) was an English computer scientist, mathematician, logician, cryptanalyst ([krɪpˈtænəlɪst]n.密码专家) and theoretical biologist. He was highly influential ([ˌɪnfluˈenʃəl]adj.有影响的) in the development of theoretical computer science, providing a formalization of the concepts of algorithm and computation with the Turing machine, which can be considered a model of a general purpose computer. Turing is widely considered to be the father of theoretical computer science and artificial intelligence.
② Real-time data (RTD) is information that is delivered immediately after collection. There is no delay in the timeliness of the information provided.
③ Predictive modeling uses statistics to predict outcomes. Most often the event one wants to predict is in the future, but predictive modelling can be applied to (适用于,应用于) any type of unknown event, regardless of when it occurred.

- A number of homes being provided with smart energy meters to become aware of energy consumption and reduce energy usage;
- Solar power garbage compactors, car recharging stations and energy saving lamps.

Some major fields of intelligent city activation are:

Innovation economy	Urban infrastructure	Governance
Innovation in industries, clusters, districts of a city	Transport	Administration services to the citizen
Knowledge workforce: Education and employment	Energy / Utilities	Participatory and direct democracy
Creation of knowledge-intensive companies	Protection of the environment / Safety	Services to the citizen: Quality of life

3. Platforms and technologies

New Internet technologies promote cloud-based services, the Internet of Things (IoT), real-world user interfaces, use of smart phones and smart meters, networks of sensors and RFIDs, and more accurate communication based on the semantic web①. They open new ways to collective action and collaborative problem solving.

Online collaborative sensor data management platforms are on-line database services that allow sensor owners to register and connect their devices to feed data into an on-line database for storage and allow developers to connect to the database and build their own applications based on that data.

The city of Santander in Cantabria, northern Spain, has 20,000 sensors connecting buildings, infrastructure, transport, networks and utilities. It offers a physical space for experimentation and validation of the IoT functions, such as interaction and management protocols, device technologies, and support services such as discovery, identity management② and security. In Santander, the sensors monitor the levels of pollution, noise, traffic and parking.

Electronic cards (known as smart cards③) are another common platform in smart city contexts. These cards possess a unique encrypted identifier that allows the owner to log in to a range of government provided services (or e-services) without setting up multiple accounts. The single identifier allows governments to aggregate data about citizens and their preferences to

---

① The Semantic Web is an extension of the Web through standards by the World Wide Web Consortium (W3C). The standards promote common data formats and exchange protocols on the Web, most fundamentally the Resource Description Framework (RDF).

② In computer security, Identity and Access Management (IAM) is the security and business discipline that "enables the right individuals to access the right resources at the right times and for the right reasons". It addresses the need to ensure appropriate access to resources across increasingly heterogeneous ([ˌhetərəu'dʒiːniəs] adj.不同种类的,异类的) technology environments and to meet increasingly rigorous (['rigərəs] adj.严格的, 严厉的, 严峻的) compliance requirements.

③ A smart card, chip card, or integrated circuit card (ICC) is any pocket-sized card that has embedded Integrated Circuits (IC,集成电路).

improve the provision of services and to determine common interests of groups. This technology has been implemented in Southampton.

4. Smart City Roadmap

A smart city roadmap consists of four/three (the first is a preliminary check) major components:

(1) Define exactly what is the community: maybe that definition can condition what you are doing in the subsequent steps; it relates to geography, links between cities and countryside and flows of people between them; maybe, even, that in some countries the definition of city/community that is stated does not correspond effectively to what, in fact, happens in the real life.

(2) Study the community: Before deciding to build a smart city, first we need to know why. This can be done by determining the benefits of such an initiative. Study the community to know the citizens, the business's needs, know the citizens and the community's unique attributes, such as the age of the citizens, their education, hobbies, and attractions of the city.

(3) Develop a smart city policy: Develop a policy to drive the initiatives, where roles, responsibilities, objective, and goals, can be defined. Create plans and strategies on how the goals will be achieved.

(4) Engage the citizens: This can be done by engaging the citizens through the use of e-government① initiatives, open data②, sport events, etc.

In short, People, Processes, and Technology are the three principles of the success of a smart city initiative. Cities must study their citizens and communities, know the processes, business drivers, create policies, and objectives to meet the citizens' needs. Then, technology can be implemented to meet the citizens' need in order to improve the quality of life and create real economic opportunities. This requires a holistic customized approach that accounts for city cultures, long-term city planning, and local regulations.

New Words

community	[kə'mju:niti]	n.	公社，团体，社会
inefficiency	[ˌini'fiʃənsi]	n.	无效率，无能
consumption	[kən'sʌmpʃən]	n.	消费，消费量
challenge	['tʃælindʒ]	n.	挑战
		vt.	向……挑战

---

① E-government (short for electronic government) is the use of electronic communications devices, computers and the Internet to provide public services to citizens.
② Open data is the idea that some data should be freely available to everyone to use and republish as they wish, without restrictions from copyright, patents or other mechanisms of control. The goals of the open data movement are similar to those of other "open" movements such as <u>open source</u> (开源), open hardware, open content, and open access.

unclear	[ʌnˈkliə]	adj. 不清楚的
cyberville	[ˈsaibəvil]	n. 网络城市
interpretation	[inˌtəːpriˈteiʃən]	n. 解释，阐明
flexicity	[fleksiˈsiti]	n. 弹性城市，柔性城市
agriculture	[ˈægrikʌltʃə]	n. 农业，农艺，农学
restructure	[riˈstrʌktʃə]	vt. 更改结构，重建构造，调整，改组
online	[ˈɔnlain]	n. 联机，在线式
constant	[ˈkɔnstənt]	n. 不变的，持续的，坚决的
agenda	[əˈdʒendə]	n. 议程
distil	[diˈstil]	vt. 提取，精炼，浓缩；融入
embedding	[emˈbediŋ]	n. 嵌入，埋入
territorialisation	[ˈteritɔːriəlaiˈzeiʃən]	n. 按地区分配
utilise	[ˈjuːtilaiz]	vt. 利用
positively	[ˈpɔzətivli]	adv. 断然地，肯定地
impact	[ˈimpækt]	n. 冲击，影响
		vt. 对……发生影响
e-participation	[iː-pɑːˌtisiˈpeiʃən]	n. 电子分享，电子参与
participation	[pɑːˌtisiˈpeiʃən]	n. 分享，参与
co-design	[ˈkəu-diˈzain]	v. 协作设计，合作计划
adapt	[əˈdæpt]	vt. 使适应；改编
dimension	[diˈmenʃən]	n. 维(数)，度(数)，元
nerve	[nəːv]	n. 神经
organ	[ˈɔːgən]	n. 元件，机构
collaboration	[kəˌlæbəˈreiʃən]	n. 协作
experimental	[eksˌperiˈmentl]	adj. 实验的，根据实验的
instrumentation	[ˌinstrumenˈteiʃən]	n. 使用仪器
controversy	[ˈkɔntrəvəːsi]	n. 论争，辩论，论战
surveillance	[səːˈveiləns]	n. 监视，监督
garbage	[ˈgɑːbidʒ]	n. 垃圾，废物
compactor	[kəmˈpæktə]	n. 垃圾捣碎机，压土机，夯土机
recharge	[ˈriːtʃɑːdʒ]	vt. 再充电
activation	[ˌæktiˈveiʃən]	n. 激活，活动
participatory	[pɑːˈtisipeitəri]	adj. 供人分享的
knowledge-intensive	[ˈnɔlidʒinˈtensiv]	adj. 知识密集型的
semantic	[siˈmæntik]	adj. 语义的
collective	[kəˈlektiv]	adj. 集体的
		n. 集体
collaborative	[kəˈlæbəreitiv]	adj. 合作的，协作的，协力完成的
pollution	[pəˈluːʃən]	n. 污染

possess	[pəˈzes]	vt. 拥有，持有，支配
unique	[juːˈniːk]	adj. 唯一的，独特的
roadmap	[ˈrəudmæp]	n. 路标
preliminary	[priˈliminəri]	adj. 预备的，初步的
subsequent	[ˈsʌbsikwənt]	adj. 后来的，并发的
countryside	[ˈkʌntrisaid]	n. 乡下地方，乡下居民
initiative	[iˈniʃiətiv]	n. 倡议；主动性；主动精神
		adj. 自发的，创始的；初步的
hobby	[ˈhɔbi]	n. 业余爱好
attraction	[əˈtrækʃən]	n. 吸引，吸引力，吸引人的事物
holistic	[həuˈlistik]	adj. 整体的，全盘的

📖 Phrases

smart city	智慧城市
information systems	信息系统
power plant	发电厂，发电站
water supply	给水，自来水，供水系统
waste management	废物管理
real-time monitoring system	实时监控系统
reduce cost	降低成本
digital city	数字城市
electronic community	电子社区
intelligent city	智能城市
health care	卫生保健
climate change	气候变化
economic restructuring	经济改革
metropolitan city-region	都市圈
in a manner	在某种意义上；在一定程度上；以……方式
business cluster	产业聚集
urban agglomeration	都市聚集
physical infrastructure	物质基础设施
physical assets	实物资产
data analytics	数据分析
open innovation	开放式创新
human intelligence	人工智能
collective intelligence	群体智能

embedded intelligence	嵌入智能
cognitive competence	认知能力，感知能力
orchestration intelligence	业务流程智能化
empowerment intelligence	赋权智能
instrumentation intelligence	仪器仪表智能化
predictive modelling	预测模型
solar power	太阳能
car recharging station	汽车充电站
energy saving lamp	节能灯
cloud-based service	基于云的服务
user interface	用户界面，用户接口
smart meter	智能仪表
identity management	身份管理
smart card	智能卡
real life	现实生活，实际生活

Abbreviations

ICT (Information and Communication Technology)	信息和通信技术
EU (European Union)	欧盟
IP (Internet Protocol)	网际协议

Exercises

I. Answer the following questions according to the text.

1. What is a smart city?
2. What do the city's assets include?
3. What is the goal of building a smart city?
4. What does ICT stand for? What is it used to do?
5. What are the four factors listed by Deakin and Al Wear that contribute to the definition of a smart city?
6. What are the three ways in which the forms of intelligence in smart cities have been demonstrated?
7. How is instrumentation intelligence implemented in Amsterdam?
8. What are online collaborative sensor data management platforms?
9. What are electronic cards known as? What do they possess?
10. What are the major components a smart city roadmap consists of?

II. Translate the following terms or phrases from English into Chinese and vice versa.

1. consumption
2. collaboration
3. embedding
4. n.分享，参与
5. cyberville

6. inefficiency
7. knowledge-intensive
8. online
9. vt.更改结构，重建构造，调整
10. flexicity

1. _____
2. _____
3. _____
4. _____
5. _____

6. _____
7. _____
8. _____
9. _____
10. _____

III. Fill in the blanks with the words given below.

emergence   move   advanced   experience   reduced
microprocessors   smart   security   networking   risk

The Right Security for the Internet of Things

The Internet of Things (IoT) is moving from a centralized structure to a complex network of decentralized smart devices. This shift promises entirely new services and business opportunities. An increasingly connected world will see the growing __1__ and cloud-enablement of all sorts of physical devices from machines through cars to home appliances. It is even transforming manufacturing as we __2__ towards the fourth "industrial revolution". Accelerated by initiatives such as the German Government's Industrie 4.0 (industrial internet) project, this stage of industrial development will see the __3__ of smart factories powered and secured by smart semiconductor solutions capable of sharing information and optimizing processes across the entire value chain.

1. Security matters

The IoT is built on many different semiconductor technologies, including power management devices, sensors and __4__. Performance and security requirements vary considerably from one application to another. One thing is constant, however. And that is the fact that the success of __5__ homes, connected cars and Industrie 4.0 factories hinges on user confidence in robust, easy-to-use, fail-safe security capabilities. The greater the volume of sensitive data we transfer over the IoT, the greater the __6__ of data and identity theft, device manipulation, data falsification, IP theft and even server/network manipulation.

2. Partner of choice for IoT security

Infineon has developed a broad range of easy-to-deploy semiconductor technologies to counter growing __7__ threats in the IoT. These solutions enable system and device manufacturers as well as service providers to capitalize on growth opportunities by integrating the right level of security without compromising on the user __8__. Complemented by software and supporting services, our hardware-based products create an anchor of trust for security implementations, supporting device integrity checks, authentication and secure key management.

3. Broad portfolio of security solutions

For almost 30 years, we have been providing security solutions to protect users' data and have already shipped nearly 20 billion security controllers worldwide. We are now bringing our market and innovation leadership to the IoT realm, supporting new, more sustainable ways of communicating and living. Our portfolio ranges from basic authentication products (OPTIGA™ Trust) to __9__ implementations (OPTIGA™ TPM, OPTIGA™ Trust P and OPTIGA™ Trust E) protecting integrity, authenticity and confidentiality of information to enable security in the IoT. Further highlights include M2M, Fido, boosted NFC, USB, RFID and My-d™ as well as CIPURSE™ innovations.

Across the full application spectrum, you can rely on us to deliver:
• Value through greater differentiation at __10__ cost to keep you one step ahead;
• Ease of deployment with no loss of user experience;
• Tailored, granular security functionality, balancing performance with cost;
• Reliability with high-quality products and a proven track record to reduce uncertainty in new business models.

IV. Translate the following passages from English to Chinese.

Machine to machine refers to direct communication between devices using any communications channel, including wired and wireless. Machine to machine communication can include industrial instrumentation, enabling a sensor or meter to communicate the data it records (such as temperature, inventory level, etc.) to application software that can use it (for example, adjusting an industrial process based on temperature or placing orders to replenish inventory). Such communication was originally accomplished by having a remote network of machines relay information back to a central hub for analysis, which would then be rerouted into a system like a personal computer.

More recent machine to machine communication has changed into a system of networks that transmits data to personal appliances. The expansion of IP networks around the world has made machine to machine communication quicker and easier while using less power. These networks also allow new business opportunities for consumers and suppliers.

Text B

The Impact of the Internet of Things on Big Data

The Internet of Things (IoT) is on its way to becoming the next technological revolution. According to Gartner, revenue generated from IoT products and services will <u>exceed</u> $300 billion in 2020, and that probably is just the tip of the iceberg. Given the massive amount of revenue and data that the IoT will generate, its impact will be felt across the entire big data universe. It will force companies to upgrade current tools and processes as well as the technology and to accommodate this additional data volume and take advantage of the insights all this new data undoubtedly will deliver.

Let's take a closer look at the various ways in which the IoT will impact big data.

1. Data Storage

When we talk about IoT, one of the first things that comes to mind is a huge, continuous stream of data hitting companies' data storage. Data centers must be equipped to handle this additional load of heterogeneous data.

In response to this direct impact on big data storage infrastructure, many organizations are moving toward the Platform as a Service (PaaS)① model instead of keeping their own storage infrastructure, which would require continuous expansion to handle the load of big data. PaaS is a cloud-based, managed solution that provides scalability, flexibility, compliance, and a sophisticated architecture to store valuable IoT data.

Cloud storage options include private, public, and hybrid models. If companies have sensitive data or data that is subject to regulatory compliance requirements that require heightened security, a private cloud model might be the best fit. Otherwise, a public or hybrid model can be chosen as storage for IoT data.

2. Big Data Technologies

When selecting the technology stack for big data processing, the tremendous influx of data

---

① Platform as a service (PaaS) is a category of cloud computing services that provides a platform allowing customers to develop, run, and manage applications without the complexity of building and maintaining the infrastructure typically associated with developing and launching an app. PaaS can be delivered in two ways: as a public cloud service from a provider, where the consumer controls software deployment with minimal configuration options, and the provider provides the networks, servers, storage, OS, 'middleware' (['mɪdlweə]n.中间件) (i.e.; java runtime, .net runtime, integration, etc.), database and other services to host the consumer's application; or as a private service (software or appliance) inside the firewall, or as software deployed on a public infrastructure as a service.

that the IoT will deliver must be kept in mind. Organizations will have to adapt technologies to map with IoT data. Network, disk, and compute power all will be impacted and should be planned to take care of this new type of data.

From a technology perspective, the most important thing is to receive events from IoT-connected devices. The devices can be connected to the network using WiFi, Bluetooth, or another technology, but must be able to send messages to a broker using some well-defined protocol. One of the most popular and widely used protocols is Message Queue Telemetry Transport (MQTT)①. Mosquitto② is a popular open-source MQTT broker.

Once the data is received, the next consideration is the technology platform to store the IoT data. Many companies use Hadoop③ and Hive to store big data. But for IoT data, NoSQL document databases like Apache CouchDB④ are more suitable because they offer high throughput and very low latency. These types of databases are schema-less, which supports the flexibility to add new event types easily. Other popular IoT tools are Apache Kafka⑤ for intermediate message brokering and Apache Storm⑥ for real-time stream processing.

3. Data Security

The types of devices that make up the IoT and the data they generate will vary in nature—raw devices, varied types of data, and communication protocol — and this carries inherent data security risks. This heterogeneous IoT world is new to security professionals, and that lack of experience increases security risks. Any attack could threaten more than just the data – it also could damage the connected devices themselves.

IoT data will require organizations to make some fundamental changes to their security landscape. As the IoT evolves, an unmanaged number of IoT devices will be connected to the network. These devices will be of different shapes and sizes and located outside the network,

---

① MQTT is an ISO standard (ISO/IEC PRF 20922) publish-subscribe-based "lightweight" messaging protocol for use on top of the TCP/IP protocol. The publish-subscribe messaging pattern requires a message broker. The broker is responsible for distributing messages to interested clients based on the topic of a message.

② Eclipse Mosquitto™ is an open source (EPL/EDL licensed) message broker that implements the MQTT protocol versions 3.1 and 3.1.1. MQTT provides a lightweight method of carrying out messaging using a publish/subscribe model. This makes it suitable for (适合……的) "Internet of Things" messaging such as with low power sensors or mobile devices such as phones, embedded computers or microcontrollers like the Arduino.

③ Apache Hadoop is an open-source software framework for distributed storage and distributed processing of very large data sets on computer clusters built from commodity hardware. All the modules in Hadoop are designed with a fundamental assumption that hardware failures are common and should be automatically handled by the framework.

④ Apache CouchDB is open source database software that focuses on ease of use and having an architecture that "completely embraces the Web". It has a document-oriented (面向文档的) NoSQL database architecture and is implemented in the concurrency-oriented (面向并发的) language Erlang; it uses JSON to store data, JavaScript as its query language using MapReduce, and HTTP for an API.

⑤ Apache Kafka is an open-source message broker project developed by the Apache Software Foundation written in Scala. The project aims to provide a unified (['juːnifaid] 统一的, 统一标准的, 一元化的), high throughput, low latency (['leitənsi]n.反应时间) platform for handling real-time data feeds.

⑥ Apache Storm is a distributed stream processing computation framework written predominantly in the Clojure programming language.

capable of communicating with corporate applications. Therefore, each device should have a non-repudiable identification for authentication purposes. Enterprises should be able to get all the details about these connected devices and store them for audit purposes. All internal and external core routers/switches should be instrumented with X.509 certificates[①] for creating trusted connectivity between public and private networks.

A multi-layered security system and proper network segmentation will help prevent attacks and keep them from spreading to other parts of the network. A properly configured IoT system should follow fine-grained network access control policies to check which IoT devices are allowed to connect. Software-defined networking (SDN)[②] technologies, in combination with network identity and access policies, should be used to create dynamic network segmentation. SDN-based network segmentation also should be used for point-to-point and point-to-multipoint encryption based on some SDN/PKI amalgamation.

4. Big Data Analytics

IoT and big data basically are two sides of the same coin. Managing and extracting value from IoT data is the biggest challenge that companies face. Organizations should set up a proper analytics platform/infrastructure to analyze the IoT data. And they should remember that not all IoT data is important.

A proper analytics platform should be based on three parameters: performance, right-size infrastructure, and future growth. For performance, a bare-metal server, a single-tenant physical server dedicated to a single customer, is the best fit. For infrastructure and future growth, hybrid is the best approach. Hybrid deployments, which consist of cloud, managed hosting, colocation, and dedicated hosting, combine the best features from multiple platforms into a single optimal environment. Managed Service Providers (MSPs) are also working on their platforms to handle IoT data. MSP vendors are typically working on the infrastructure, performance, and tools side to cover the entire IoT domain.

An IoT device generates continuous streams of data in a scalable way, and companies must handle the high volume of stream data and perform actions on that data. The actions can be event correlation, metric calculation, statistics preparation, and analytics. In a normal big data scenario, the data is not always stream data, and the actions are different. Building an analytics solution to manage the scale of IoT data should be done with these differences in mind.

The growth of the IoT heralds a new age of technology, and organizations that wish to

---

① In cryptography, X.509 is an important standard for a public key infrastructure (PKI, 公钥基础设施) to manage digital certificates and public-key encryption and a key part of the Transport Layer Security protocol used to secure web and email communication.

② Software-defined networking (SDN) is an approach to computer networking that allows network administrators to manage network services through abstraction ([æb'strækʃən] n.提取) of lower-level functionality. SDN is meant to (有意要, 打算) address the fact that the static architecture of traditional networks doesn't support the dynamic, scalable (['skeiləbl] adj.可升级的) computing and storage needs of more modern computing environments such as data centers.

participate in this new era will have to change the way they do things to accommodate new data types and data sources. And these changes likely are just the beginning. As the IoT grows and businesses grow with IoT, they will have many more challenges to solve.

New Words

exceed	[ik'si:d]	vt. 超越，胜过
		vi. 超过其他
generate	['dʒenə,reit]	vt. 产生，发生
universe	['ju:nivə:s]	n. 宇宙，世界，万物，领域
process	[prə'ses]	n. 过程，方法，程序，步骤
		vt. 加工，处理
insight	['insait]	n. 洞察力，见识
undoubtedly	[ʌn'dautidli]	adv. 毋庸置疑地，的确地
huge	[hju:dʒ]	adj. 巨大的，极大的，无限的
continuous	[kən'tinjuəs]	adj. 连续的，持续的
stream	[stri:m]	n. 流，一串
		v. 流，涌
heterogeneous	[,hetərəu'dʒi:niəs]	adj. 不同种类的，异类的
compliance	[kəm'plaiəns]	n. 依从，顺从
sophisticated	[sə'fistikeitid]	adj. 复杂的，先进的；老练的，富有经验的
valuable	['væljuəbl]	adj. 有价值的
hybrid	['haibrid]	n. 混合物
		adj. 混合的
sensitive	['sensitiv]	adj. 敏感的，灵敏的
regulatory	['regjulətəri]	adj. 调整的
heighten	['haitn]	v. 提高，升高
tremendous	[tri'mendəs]	adj. 极大的，巨大的
influx	['inflʌks]	n. 流入
perspective	[pə'spektiv]	n. 观点，看法，观察
broker	['brəukə]	n. 中介，中间设备
consideration	[kənsidə'reiʃən]	n. 考虑，需要考虑的事项
suitable	['sju:təbl]	adj. 适当的，相配的
throughput	['θru:put]	n. 生产量，生产能力，吞吐量
intermediate	[,intə'mi:djət]	adj. 中间的
		n. 媒介
raw	[rɔ:]	adj. 未加工的，处于自然状态的
threaten	['θretn]	vt. 恐吓，威胁
non-repudiable	[nɔŋ-ri'pju:diəbl]	adj. 不可否认的

switch	[swɪtʃ]	n.	交换机
instrument	[ˈɪnstrəmənt]	n.	工具，器械，器具，手段
amalgamation	[əˌmælgəˈmeɪʃən]	n.	融合，合并
bare-metal	[bɛə-ˈmetl]	n.	裸机
colocation	[kəʊləʊˈkeɪʃən]	n.	场地出租
correlation	[ˌkɔriˈleɪʃən]	n.	相关(性)

Phrases

big data	大数据
tip of the iceberg	露出水面的冰山顶，事物的表面部分
in response to	响应，适应
be kept in mind	牢记在心
real-time stream processing	实时流处理
in nature	实际上，本质上
multi-layered security system	多层安全系统
network segmentation	网段
in combination with …	与……结合
stream data	流数据

Abbreviations

PaaS (Platform as a Service)	平台即服务
MQTT (Message Queue Telemetry Transport)	消息队列遥测传输
SDN (Software-Defined Networking)	软件定义网络
PKI (Public Key Infrastructure)	公钥基础设施
MSPs (Managed Service Providers)	托管服务提供者

Exercises

I. Answer the following questions according to the text.

1. What is one of the first things that comes to mind when we talk about IoT?

2. What are many organizations are moving toward to respond to the direct impact on big data storage infrastructure?

3. What is PaaS?

4. What are the three models cloud storage have?

5. When selecting the technology stack for big data processing, what must be kept in mind?

6. What is one of the most popular and widely used protocols?

7. Why are NoSQL document databases like Apache CouchDB more suitable for IoT data?

8. What will help prevent attacks and keep them from spreading to other parts of the

network?

    9. What is the biggest challenge that companies face? What should they do?

    10. What are MSP vendors typically working on?

II. Translate the following terms or phrases from English into Chinese and vice versa.

1. bare-metal
2. continuous
3. heterogeneous
4. n.生产量，生产能力
5. sophisticated
6. hybrid
7. suitable
8. threaten
9. valuable
10. n.过程，方法，程序，步骤

Text A 参考译文

智 慧 城 市

    智慧城市是一种城市发展愿景，它以安全的方式整合多个信息和通信技术(ICT)及物联网(IoT)解决方案，以便管理城市的资产。城市的资产包括但不限于地方部门的信息系统、学校、图书馆、交通系统、医院、电厂、供水网络、废物管理以及执法和其他社区服务。建设智慧城市的目标是通过使用城市信息和技术提高生活质量、增强服务效率和满足居民的需求。ICT 允许城市官员直接与社区和城市基础设施进行交互，并监测城市的状况与变化，以及如何实现更好的生活质量。通过使用与实时监控系统集成的传感器，从公民和设备那里收集数据，然后进行处理和分析。收集信息和知识是解决效率低下的关键。

    ICT 用于提高城市服务的质量、性能和互动性，降低成本和资源消耗，并改善公民与政府之间的联系。开发智慧城市应用程序的目的是改进城市流的管理，并允许实时应对挑战。因此，一个智慧城市可能更能够应对挑战，而不只是一个与其公民保持简单的"交易"关系的城市。然而，这个术语本身的细节仍然不够清晰，因此，人们对此有许多解释。

    已经用于类似概念的其他术语包括网络城市、数字城市、电子社区、弹性城市、信息城市、智能城市、基于知识的城市和泛在城市。

    发展智慧城市技术的领域包括政府服务、运输和交通管理、能源、保健、水、创新的城市农业及废物管理。

    主要技术、经济和环境变化已经引起了人们对智慧城市的兴趣，这些变化包括气候变化、经济重组、网络零售和娱乐，城市人口增长和公共财政压力。欧盟(EU)一直致力于增加其大都市圈实现"智能"的城市增长策略。欧盟根据"欧洲数字议程"制订了一系列计划。2010 年，它强调其重点是加强 ICT 服务的创新和投资，以改善公共服务和生活质量。据 Arup 估计，到 2020 年用于智慧城市服务的全球市场每年将达到 4000 亿美元。

1. 术语

由于在智慧城市标签下已经实施了众多技术，难以精确定义智慧城市。Deakin 和 Al Wear 列出了有助于确定智慧城市的四个因素：
- 将电子和数字技术广泛应用于社区和城市；
- 利用信息通信技术改变区域内的生活和工作环境；
- 将这些 ICT 纳入政府系统；
- 整合 ICT 和人以增强其区域创新与知识。

Deakin 将智慧城市定义为利用 ICT 来满足市场(城市公民)的需求，并且社区必须参与智慧城市的构建。因此，智慧城市不仅具有特定领域的 ICT 技术，而且已经实施这些技术并在一定程度上对当地社区产生积极影响。

2. 特性

有人认为，智慧城市(社区、商业集群、城市群或区域)使用信息技术应达到以下目的：

(1) 通过人工智能和数据分析更有效地利用物理基础设施(道路、建筑环境和其他有形资产)，以支持强大和健康的经济、社会及文化发展。

(2) 通过开放式创新流程和电子参与，使当地居民有效地参与地方治理和决策，通过电子政务提高城市机构的集体智慧，重点是公民参与和共同设计。

(3) 通过学习、适应和创新提高城市的智能，更有效和及时地应对不断变化的情况。

它们的演变方向为有力整合人类智力、集体智慧以及城市内人工智能的各个方面。城市的智能"在于数字电信网络(神经)、无处不在地嵌入智能(大脑)、传感器和标签(感觉器官)和软件(知识和认知能力)越来越有效的结合"。

智慧城市中的智能形式有以下三种展示方式：

(1) 业务流程智能化：城市建立解决问题和协作的机构与社区。例如在布莱奇利公园，那里的 Nazi Enigma 密码由一个叫阿兰图灵领导的团队解码。这被称为智慧城市或智能社区的第一个例子。

(2) 赋权智能：城市提供开放平台、实验设施和智能城市基础设施，以便在某些地区集群创新。这在斯德哥尔摩的 Kista 科学城和香港的数码港区都能看到。在墨尔本也建立了类似的设施。

(3) 仪器智能：通过收集整个城市地区的实时数据、分析和运行预测模型来实现城市基础设施智能化。围绕这一点有很多争议，特别是关于智慧城市的监控问题。仪表智能的例子已经在阿姆斯特丹实现。它通过以下方式实现：
- 开放给研究人员开发应用程序的公共 IP 基础设施；
- 无线仪表和设备按时定点发送信息；
- 许多家庭提供智能电表，以了解能量消耗和减少能源使用；
- 使用太阳能发电的垃圾压实机、汽车充电站和节能灯。

智慧城市激活了以下主要领域：

创新经济	城市基础设施	治理
行业创新、集群、城市区域	运输	对公民的管理服务
知识劳动力：教育和就业	能源/公用事业	参与式和直接民主
创建知识密集型公司	环境保护/安全	对公民的服务：生活质量

3. 平台和技术

新的因特网技术促进了基于云的服务、物联网(IoT)、真实世界的用户界面、智能手机和智能仪表的使用、传感器和 RFID 的网络以及基于语义网络的更准确的通信。它们打开了新的集体协作行动和合作解决问题的方法。

在线协作传感器数据管理平台是在线数据库服务，其允许传感器所有者注册并连接他们的设备以将数据馈送到用于存储的在线数据库，并允许开发者连接到该数据库，且基于该数据来构建他们自己的应用。

西班牙北部坎塔布里亚的桑坦德市有 2 万个传感器，连接了建筑物、基础设施、交通、网络和公用设施。它为物联网功能的实验和验证提供了物理空间，例如交互和管理协议、设备技术，并且支持如发现、身份管理和安全这类服务。在桑坦德，传感器监测污染、噪声、交通和停车程度。

电子卡(称为智能卡)是智慧城市环境中的另一个通用平台。这些卡具有唯一的加密标识符，允许所有者登录到一系列政府提供的服务(或电子服务)而无需设置多个账户。单一标识符允许政府汇总关于公民及其偏好的数据以改善服务，并确定群体的共同利益。这项技术已在南安普敦实施。

4. 智慧城市路线图

一个智慧城市路线图由四/三个(第一个是初步检查)主要部分组成：

(1) 确定社区是什么：也许这个定义可以限制你在后续步骤中做什么；它涉及地理、城市和农村之间的联系以及它们之间的人员流动。也许在一些国家，这里所陈述的城市/社区的定义与现实生活不相符。

(2) 研究社区：在决定建设智慧城市之前，首先我们需要知道其理由。这可以通过确定这种举措的好处来实现。研究社区以了解公民和企业的需要，了解公民和社区的独特属性，如公民的年龄、他们受教育的程度、爱好和城市的景点。

(3) 制定智慧城市政策：制定政策以推动各项举措，可以定义角色、责任、目标和目的。制定实现目标的计划和战略。

(4) 公民参与：可以通过使用电子政府举措、开放数据及体育赛事等方式吸引公民。

总之，人、过程和技术是智慧城市计划成功的三个原则。城市必须研究其公民和社区，了解过程和业务驱动因素，制定政策和目标以满足公民的需要。然后，可以实施技术以满足公民的需要，以提高生活质量并创造真正的经济机会。这需要一个全面的个性化方法，包括建立城市文化，制定长期城市规划和地方法规。

附录 I 自 测 题

一、根据英文单词,写出中文意思(20×0.5=10,共10分)

英文单词	中文意思
receiver	
self-explanatory	
framework	
broadband	
match	
monolithic	
incompatible	
interaction	
scalable	
handle	
browser	
interface	
redundancy	
platform	
spectral	
addressability	
barcode	
configure	
functional	
identity	

二、根据中文意思，写出英文单词(20×0.5=10，共10分)

中 文 意 思	英 文 单 词
vt. 感知，感到，认识	
n. 传感器	
n. 标签，标识	
v. 振荡	
adj. 可追踪的，起源于	
n. 会话，交谈	
n. 光纤	
n. 超文本	
n. 数据包	
n. 路由器	
n. 模块	
n. 标识符	
v. 封装	
n. 适配器	
n. 扫描器	
n. 碰撞，冲突	
vt. 刷……卡	
vt. 排除，消除	
adj. 红外线的 n. 红外线	
n. 拓扑，布局	

三、根据英文词组，写出中文意思(30×0.5=15，共15分)

词 组	中 文 意 思
at full capacity	
data processing	
interface software	
storage space	

词　　　组	中　文　意　思
spectral efficiency	
channel bandwidth	
global roaming	
address space	
autonomous control	
data capture	
event transfer	
Internet of Things	
Internet Protocol (IP)	
parallel computing	
radio tag	
trigger action	
bar code	
smart dust	
set of rules	
bus network	
gateway server	
graphical user interface	
layer-3 switch	
network bridge	
physical address	
remote access	
pattern recognition	
tamper resistance	
transport protocol	
margin for error	

四、根据英文缩写，写出英文完整形式及中文意思(10×2＝20，共 20 分)

缩　　写	英文完整形式	中　文　意　思
ATM		
CMOS		
CSMA		
DNS		
FTP		
IC		
VPDN		
WiFi		
XML		
EEPROM		

五、翻译句子(10×1.5＝15，共 15 分)

(1) The Internet of Things is a network of Internet-enabled objects, together with web services that interact with these objects.

(2) You might create an ad hoc network with many readers monitoring hundreds of tag sensors spread out across tens of square miles.

(3) The latest Japanese vacuum cleaners contain sensors that detect the amount of dust and type of floor.

(4) Domain administrator can disable or delete the computer account for this server.

(5) Using radio means that the tag no longer has to be visible on the object to which it is attached, the tag can be hidden inside the item or box that is to be identified and still be read.

六、把下列句子翻译为英文(5×2＝10，共 10 分)

(1) 为什么人人都需要一个新的协议以便在网络上实现管理信息的交流？
(2) 因特网让计算机用户与全球的计算机连接。
(3) 像零售商和医院这样的大型机构都在应用 RFID 标签来管理库存。
(4) 现在一个服务器可以容易地实现电脑、储存设备和路由器之间的交换。
(5) 在互联网上每一个机器有独特的识别号码，称为 IP 地址。

七、根据下列方框中所给的词填空(10×1＝10，共 10 分)

A　access	B　devices	C　identifier	D　services	E　broadband
F　radio	G　hotspots	H　network	I　Provider	J　protocol

1. Internet

A global ___1___ connecting millions of computers. More than 100 countries are linked into exchanges of data, news and opinions. Unlike online services, which are centrally controlled, the Internet is decentralized by design. Each Internet computer, called a host, is independent. Its operators can choose which Internet services to use and which local ___2___ to make available to the global Internet community. Remarkably, this anarchy by design works exceedingly well. There are a variety of ways to ___3___ the Internet. Most online services, such as America Online, offer access to some Internet services. It is also possible to gain access through a commercial Internet Service ___4___ (ISP).

2. IP address

An ___5___ for a computer or device on a TCP/IP network. Networks using the TCP/IP ___6___ route messages based on the IP address of the destination. The format of an IP address is a 32-bit numeric address written as four numbers separated by periods. Each number can be zero to 255. For example, "1.160.10.240" could be an IP address.

Within an isolated network, you can assign IP addresses at random as long as each one is unique. However, connecting a private network to the Internet requires using registered IP addresses (called Internet addresses) to avoid duplicates.

3. Wireless Internet

Wireless Internet enables wireless connectivity to the Internet via ___7___ waves rather than wires on a person's home computer, laptop, smartphone or similar mobile device. Wireless Internet can be accessed directly through providers like AT&T, Verizon, T-Mobile, Boingo and Clearwire.

While most wireless Internet options lack the high speed of landline ___8___ Internet connections such as cable and DSL, newer wireless Internet technologies like EV-DO and WiMAX are narrowing the gap, with maximum speeds of up to 7 Mbps in some cases.

WiFi ___9___ and wireless LANs are also options for wireless Internet connectivity. In these cases, Internet connectivity is typically delivered to a network hub via a wired connection like satellite, cable, DSL or fiber optics and then made available to wireless ___10___ via a wireless access point.

八、根据下列短文回答问题，回答使用英文(5×2 = 10，共 10 分)

◇◇◇　The future of IoT　◇◇◇

When looking at today's state of the art technologies, they should give a clear indication of how the Internet of Things will be implemented on a universal level in the years to come as well as indicate important aspects that need to be further studied and developed in the coming years. Firstly, the need exists for significant work in the area of governance. Without a standardised approach it is likely that a proliferation of architectures, identification schemes, protocols and frequencies will develop side by side, each one dedicated to a particular and separate use. This will inevitably lead to a fragmentation of the IoT, which could hamper its popularity and become a major obstacle in its roll out. Interoperability is a necessity, and inter-tag communication is a pre-condition in order for the adoption of IoT to be wide-spread.

In the coming years, technologies necessary to achieve the ubiquitous network society are expected to enter the stage of practical use. It is widely expected that RFID technology will become mainstream in the retail industry around 2010. As this scenario will evolve, a vast amount of objects will be addressable, and could be connected to IP-based networks, to constitute the very first wave of the "Internet of Things". There will be two major challenges in order to guarantee seamless network access: the first issue relates to the fact that today different networks coexist; the other issue is related to the sheer size of the "IoT". The IT industry has no experience in developing a system in which hundreds of millions of objects are connected to IP networks. Other current issues, such as address restriction, automatic address setup, security functions such as authentication and encryption, and multicast functions to deliver voice and video signals efficiently will probably be overcome by ongoing technological developments.

Another very important aspect that needs to be addressed at this early stage is the one related to legislation. Various consumer groups have expressed strong concerns about the numerous possibilities for this technology to be misused. A clear legislative framework ensuring the right of privacy and security level for all users must therefore be implemented by all member states. A sustained information campaign highlighting the benefits of this technology to society at large must also be organised, a campaign which emphasises the benefits that this technology can bring to ordinary citizens in their every day lives be it improved food traceability, assisted living or more secure healthcare.

Traditionally, the retail and logistics industry require very low cost tags with limited features; such as an ID number and some extra user memory area, while other applications and industries will require tags that will contain a much higher quantity of data and more interactive and intelligent functions. "Data", in this context, can be seen as an "object" and under this vision a tag carries not only its own characteristics, but also the operations it can handle. The amount of intelligence that the objects in the IoT will need to have and if, how and in which cases this intelligence is distributed or centralised becomes a key factor of development in the future. As the "IQ" of "things" will grow, the pace of the development and study of behavioural requirements of these objects will also become more prevalent in order to ensure that these objects can co-exist in seamless and non-hostile environments. These developments should lead to interactive standards, followed eventually by behavioural ones.

Other topics for research include not only the integration of electronic identifiers into materials, such as ceramics, metals, or paper, but also the creation of devices from non-silicon based materials, such as, for instance, eatable tags. This will allow, for instance, embedding tags into medicines, which can be seen as a giant step to putting receivers into packaging. Additionally, the future IoT will have to be built within recyclable materials and therefore have to be fully eco-friendly. Future smart objects must also be power independent, harvesting energy from the environment in which they operate. Finally, the things of the future will have to be resistant to very harsh and extreme conditions, including temperature, vibrations, humidity, and hostile environments.

(1) What is the first important aspect that needs to be further studied and developed in the coming years?

(2) What is a pre-condition in order for the adoption of IoT to be wide-spread?

(3) What are the two major issues related to in order to guarantee seamless network access?

(4) What kind of tags the retail and logistics industry require traditionally? And what about other application and industries?

(5) What will the furture IoT have to be?

参 考 答 案

一、根据英文单词，写出中文意思(20 × 0.5 = 10，共 10 分)

英 文 单 词	中 文 意 思
receiver	n. 接收器，收信机
self-explanatory	adj. 自明的，自解释的
framework	n. 构架，框架，结构
broadband	n. 宽带
match	n. 匹配
monolithic	n. 单片电路，单块集成电路
incompatible	adj. 不兼容的；不相容的；互斥的
interaction	n. 交互作用，交感
scalable	adj. 可升级的
handle	vt. 处理，操作 n. 句柄
browser	n. 浏览器
interface	n. 接口，界面
redundancy	n. 冗余
platform	n. 平台
spectral	adj. 光谱的，频谱的
addressability	n. 可寻址能力
barcode	n. 条形码
configure	vi. 配置，设定
functional	adj. 功能的
identity	n. 身份，特性

二、根据中文意思，写出英文单词(20×0.5=10，共10分)

中 文 意 思	英 文 单 词
vt. 感知，感到，认识	sense
n. 传感器	sensor
n. 标签，标识	tag
v. 振荡	oscillate
adj. 可追踪的，起源于	traceable
n. 会话，交谈	conversation
n. 光纤	fiber
n. 超文本	hypertext
n. 数据包	packet
n. 路由器	router
n. 模块	module
n. 标识符	identifier
v. 封装	encapsulate
n. 适配器	adaptor
n. 扫描器	scanner
n. 碰撞，冲突	collision
vt. 刷……卡	swipe
vt. 排除，消除	eliminate
adj. 红外线的 n. 红外线	infrared
n. 拓扑，布局	topology

三、根据英文词组，写出中文意思(30×0.5=15，共15分)

词 组	中 文 意 思
at full capacity	满功率，满负载
data processing	数据处理
interface software	接口软件
storage space	存储空间
spectral efficiency	频谱效率，频谱利用率
channel bandwidth	频道带宽，通道带宽
global roaming	国际漫游
address space	地址空间

词 组	中 文 意 思
autonomous control	自主控制
data capture	数据捕捉
event transfer	事件传输
Internet of Things	物联网
Internet Protocol (IP)	因特网协议
parallel computing	并行计算
radio tag	无线标签，无线标识
trigger action	触发作用
bar code	条形码
smart dust	智能微尘
set of rules	规则组，规则集
bus network	总线网络
gateway server	网关服务器
graphical user interface	图形用户界面
layer-3 switch	第三层交换
network bridge	网桥
physical address	物理地址
remote access	远程访问
pattern recognition	模式识别
tamper resistance	抗干扰
transport protocol	传输协议
margin for error	误差容许量，误差限度

四、根据英文缩写，写出英文完整形式及中文意思(10×2 = 20，共20分)

缩写	英文完整形式	中 文 意 思
ATM	Asynchronous Transfer Mode	异步传输模式
CMOS	Complementary Metal Oxide Semiconductor	互补金属氧化物半导体
CSMA	Carrier Sense Multiple Access	载波侦听多路访问
DNS	Domain Name Server	域名服务器
FTP	File Transfer Protocol	文件传输协议
IC	Integrate Circuit	集成电路

缩写	英文完整形式	中文意思
VPDN	Virtual Private Dial-up Network	虚拟专用拨号网
WiFi	Wireless Fidelity	无线局域网
XML	eXtensible Markup Language	可扩展标识语言
EEPROM	Electrically Erasable Programmable Read-Only Memory	电可擦除只读存储器

五、翻译句子(10×1.5＝15，共15分)

(1) The Internet of Things is a network of Internet-enabled objects, together with web services that interact with these objects.

物联网是由联网物品构成的一个网络，其中包括与这些物品进行互动的互联网服务。

(2) You might create an ad hoc network with many readers monitoring hundreds of tag sensors spread out across tens of square miles.

你可以利用多个读取机建立一个随意网络，在数十平方公里的范围内，监测数百个标签感应器。

(3) The latest Japanese vacuum cleaners contain sensors that detect the amount of dust and type of floor.

日本最新款吸尘器装有传感器，能测出灰尘量和地板类型。

(4) Domain administrator can disable or delete the computer account for this server.

域管理员可禁用或删除此服务器上的计算机账户。

(5) Using radio means that the tag no longer has to be visible on the object to which it is attached, the tag can be hidden inside the item or box that is to be identified and still be read.

使用无线电子技术意味着商标没必要再出现在所贴物品上，因为隐藏在商品或箱里的标签仍能被识别和阅读。

六、把下列句子翻译为英文(5×2＝10，共10分)

(1) 为什么人人都需要一个新的协议以便在网络上实现管理信息的交流？

Why would anyone need a new protocol to communicate management information over a network?

(2) 因特网让计算机用户与全球的计算机连接。

The Internet is the network that allows computer users to connect with computers all over the world.

(3) 像零售商和医院这样的大型机构都在应用 RFID 标签来管理库存。

Large organizations such as retailers and hospitals are using RFID tags for managing stock levels.

(4) 现在一个服务器可以容易地实现电脑、储存设备和路由器之间的交换。

A server can now easily switch between being a computer, a storage device or a router.

(5) 在互联网上每一个机器有独特识别号码，称为 IP 地址。

Every machine on the internet has a unique identifying number, call an IP address.

七、根据下列方框中所给的词填空(10×1 = 10，共10分)

| A 3 | B 10 | C 5 | D 2 | E 8 |
| F 7 | G 9 | H 1 | I 4 | J 6 |

八、根据短文回答问题，回答请使用英文(5×2 = 10，共10分)

(1) The first aspect is the need for significant work in the area of governance.

(2) Inter-tag communication is a pre-condition in order for the adoption of IoT to be wide-spread.

(3) The first issue related to the fact that today different networks coexist; the other issue is related to the sheer size of the IoT.

(4) Traditionally, the retail and logistics industry require very low cost tags with Limited features; such as an ID number and some extra user memory area. Other applications and industries will require tags that will contain a much higher quantity of data and more interactive and intelligent functions.

(5) It will have to be built within recyclable materials and therefore have to be fully eco-friendly.

附录 II　物联网英语词汇的构成与翻译

物联网涉及计算机、通信与网络等多个方面，其专业英语新词汇呈现量多面广、含义复杂的特点。理解这些新词汇的构成并准确地用中文来阐释其意义是翻译的关键。

1. 物联网专业英语词汇的构造特点

物联网英语的新词中，新出现的、与现有词汇完全没有联系的词汇很少。绝大多数都是在现有的词汇基础上产生的，其构成方法有规律可循。

1.1　单词构成

在物联网专业英语领域中，新单词的构造方法主要有以下几种：

(1) 新赋意义：通过给某个单词赋予新的、有物联网英语特色的意义，使之成为专业单词。这是一种"旧瓶装新酒"的方法。如果不了解新单词在物联网英语中的新意义，就会造成理解的错误。例如"keying"在一般语境中为"按键"、"发报"，而在物联网专业英语中则给它赋予新的意思，即"键控"的含义。又如"carrier"的常用意思是"运送者"、"邮递员"，在物联网语境中的意思则是"载波"。又如"converter"的原意为"转换者"，在物联网中的意思则是"变频器"，是一个专业性很强的词。

(2) 复合构造：这样构造的新词来自于两个或多个现有的单词，由现有的单词组合在一起，形成一个新词。如：downlink(下行链路)、baseband(基带)、bandwidth(带宽)、crosstalk(串话)、microwave(微波)及 multiframe(多帧)等。

(3) 派生法构词：词根就是一个单词最原始的"本体"，不可以再继续拆分，是词的核心部分，词的意义主要是由它体现出来的。词根可以单独构成词，也可以彼此组合成词。词缀只能是粘附在词根上构成新词的语素，它本身不能单独构成词。派生法是英语主要的构词法，方法是通过给词根添加词缀来构成新词，这样产生的词就被称为派生词(derivative words)。加上前缀后的词，虽然意思改变，但词性保持不变。相反的，加上后缀的词，不但词义有些改变，词性也完全不同。例如，可以通过给词根加前缀"re"构造出 reset(复位)、recall(二次呼叫)；加前缀"multi"构造出 multiframe(多帧)、multipath(多径)、multiplexing(多路复接)；加前缀"un"构成 undue(未到期的)、unlimited(无限的，无约束的)、unchain(解除，释放)及 undo(撤消)等；加后缀"er"构成的词有 receiver(接收器)、router(路由器)、server(服务器)；加后缀"or"构成的词有 connector(连接器)、indicator(指示器)及 capacitor(电容器)；加后缀"ment"构成的词有 assignment(指配，分配)及 equipment(设备)等。

(4) 拼缀法构词：对原有的两个词进行剪裁，分别取其中的一部分，连成一个新词。例如，分别取出 communication(通信)和 satellite(卫星)的前三个字母 com 和 sat，拼接为 comsat(通信卫星)。又如 camcorder(摄相机)是 camera(照相机)与 recorder(录音机)的组合。

1.2 词组

词组就是许多由单词组成的固定搭配。在物联网专业英语中，词组使用广泛，其结构固定，专业意义明确。例如，data source(数据源)、data protection (数字保护)、dedicated channel(专用信道)、digital path(数字通道)、digital sensor (数字传感器)、electromagnetic coupling(电磁耦合)、encipherment scheme(加密方案)、error detection(检错)、frequency band(频带)、frame synchronization code (帧同步码)、information processing(信息处理，信息加工)及 installing test (实装测试) 均是词组。

1.3 缩略语构成

缩略语可以有效提高信息含量，更加符合网络时代有效传播的要求，因而在物联网英语中使用广泛。缩略词具有明显的行业特色，不理解缩略词往往就不能理解整段文字的意义。

(1) 首字母缩略：取各个单词的首字母组成缩略语。这种构造方法最为常见，是整个缩略词的基础和主体。如 CA(Certification Authority，认证机构)、ASP(Active Server Page，活动服务器页面)、TCP (Transmission Control Protocol，传输控制协议)、IP (Internet Protocol，因特网协议)、DNS (Domain Name Server，域名服务器)、ISP (Internet Service Provider，因特网服务提供商)以及 ATM (Asynchronous Transfer Mode，异步传输模式)均是如此构成的缩略词。

(2) 首尾字母缩略：由某个单词的第一个字母和该单词的最后一个字母(或后几个字母)缩略而成。例如：YD(yard，码)、Rd(Road，路)、ft(foot，英尺)、wt(weight，重量)等。

(3) 截割缩略：截去词的首部或尾部，留下词的其余部分构成一个新词。例如：phone(telephone，电话)、apps(applications，应用软件)及 ID(identity，身份)。

(4) 转音缩略：将某个单词转换为读音相同的另一单词，然后以数字代替，再进行缩略。例如：B2B(企业对企业)的原始形式为 Business to Business，转换为 Business two Business，再将 two 按照读音转换为数字 2，最终缩略为 B2B。类似的还有：B2C(Business-to-Consumer，企业对消费者)、C2C(Consumer-to-Consumer，消费者对消费者)等。

2. 物联网专业英语词汇的翻译方法

2.1 直译

直译就是直接译出字面意义。直译法是一种最常用的方法，特别是在翻译复合词时，大都用直译法。直译法简单易行，便于理解。例如，把 code 翻译为"码"，相关的词汇有：lead code(前导码)、line code(线路码)、linear block code(线性分组码)、nonreturn to zero code(非归零码)、parity check code(奇偶效验码)、positive and inverse code (正反码)、pseudorandom code(伪随机码)、pulse code(脉冲码)、repetition code(重复码)、telegraph code(电报电码)及 zone bit code(区位码)。又如，把"antenna"翻译为"天线"，相关的词汇有：antenna automatic tuning(天线自动调谐)、antenna directivity diagram(天线方向性图)、antenna gain(天线增益)、antenna matching device(天线匹配装置)、antenna power rating(天线额定功率)、antenna voltage rating(天线额定电压)。类似的例子有很多。

2.2 意译

当一个英语词汇没有完全对应的汉语词汇或者按照英文的字面意思直接翻译不符合汉语的表达方式时，应该意译。即通过某种转换，使用符合汉语习惯的词汇来表达其含义，而不拘谨于其字面意思。例如，将 brick-and-mortar store(水泥砖头商店)意译为"实体店"，与"网络商店"相对应。Microsoft 翻译为"微软"也属于此类。

2.3 音译

音译是指按照英语词汇的读音，将英文翻译为相应的汉语词汇。例如，将 cracker 翻译为"骇客"。因为 cracker 恶意破解商业软件、恶意入侵别人的网站，让人恐惧。如果翻译为"破解者、攻入者"，既不形象也不生动，而汉字"骇"传神地表达了人们谈之色变的恐惧感。类似的翻译还有 Email(伊妹儿，电子邮件)、OPEC(欧佩克)与 Topology(拓扑)。

公司名称及产品名称的翻译通常使用音译。例如：Google(谷歌)、Cisco(思科)、Intel(英特尔)、SIEMENS(西门子)、Nokia(诺基亚)、Amazon(亚马逊)、Wal-Mart(沃尔玛)、Xerox(施乐)及 Digi(迪进)。

2.4 音义结合法

此法兼顾语音和词义，既取其义，亦取其音。例如，把 Internet 翻译为"因特网"，取"Inter"之音、取"net"之义。

物联网的技术进步会不断涌现新的专业英语词汇，应该根据其构造特点，依据专业背景知识，运用恰当的翻译方法，用专业化语言准确与贴切地翻译，以求在实现词义的对等或等效的基础上，追求传神与精妙，以期臻于信、达、雅的境界。

附录Ⅲ 词汇总表

表中,"课次"表明了该词汇在本书中所在的单元,如 1A 标识该词在第一单元的 Text A 中出现。少量没有课次标注的是我们特意增加的词汇。这些词汇虽然在本书中没有出现,但在本行业中比较常用。

单 词 表

单词	音标	意义	课次
2D-code		n. 二维码	1A
abort	[əˈbɔːt]	vi. 异常中断,中途失败	9A
absolutely	[ˈæbsəluːtli]	adv. 完全地,绝对地	6A
abstract	[ˈæbstrækt]	n. 摘要,概要,抽象 adj. 抽象的	3A
abstraction	[æbˈstrækʃən]	n. 抽象	8A
abuse	[əˈbjuːz]	n.& v. 滥用	1B
abut	[əˈbʌt]	v. 邻接,毗邻	5A
accelerate	[ækˈseləreit]	v. 加速,促进	5A
access	[ˈækses]	n.& vt. 访问,存取	1A
accessibility	[ˌækəsesiˈbiliti]	n. 易访问的,可到达的	10B
accessible	[əkˈsesəbl]	adj. 可访问的,易接近的,可到达的	4B
accident	[ˈæksidənt]	n. 意外事件,事故	1B
accommodate	[əˈkɔmədeit]	vt. 供应,供给,使适应,调节	2A
accomplish	[əˈkɔmpliʃ]	vt. 完成,达到,实现	4B
accordingly	[əˈkɔːdiŋli]	adv. 因此,从而	1A
account	[əˈkaunt]	n. 账户	10A
accuracy	[ˈækjurəsi]	n. 精确性,正确度	4A
accurate	[ˈækjurit]	adj. 正确的,精确的	1A
acknowledgment	[əkˈnɔlidʒmənt]	n. 应答,承认	9A
acronym	[ˈækrənim]	n. 只取首字母的缩写词	6A
activate	[ˈæktiveit]	vt. 刺激,使活动 vi. 有活力	4A
activation	[ˌæktiˈveiʃən]	n. 激活,活动	11A
actuator	[ˈæktjueitə]	n. 驱动器,执行器	3B

单　　词	音　　标	意　　义	课次
adapt	[ə'dæpt]	vt. 使适应；改编	11A
adapter	[ə'dæptə]	n. 适配器	7A
adaptive	[ə'dæptiv]	adj. 适应的	3A
adaptor	[ə'dæptə]	n. 适配器	5A
address	[ə'dres]	n. 地址	2A
addressability	[ə'dresəbiliti]	n. 可寻址能力	1A
addressable	[ə'dresəbl]	adj. 可设定地址的	1A
adherence	[əd'hiərəns]	n. 依附，黏着	7B
adjust	[ə'dʒʌst]	vt. 调整，调节，校准，使适合	5B
advertiser	['ædvətaizə]	n. 登广告者，广告客户	5B
advertising	['ædvətaiziŋ]	n. 广告业，广告　adj. 广告的	4A
advisor	[əd'vaizə]	n. 顾问	1A
advocate	['ædvəkit]	vt. 提倡，主张，拥护	5A
agenda	[ə'dʒendə]	n. 议程	11A
aggregate	['ægrigeit]	v. 聚集，集合，合计	8A
agriculture	['ægrikʌltʃə]	n. 农业，农艺，农学	11A
aisle	[ail]	n. 走廊，过道	6A
alarm	[ə'lɑːm]	n. 警报，警告器　vt. 恐吓，警告	2B
algorithm	['ælgəriðəm]	n. 算法	8A
alignment	[ə'lainmənt]	n. 队列	5B
allocate	['æləukeit]	vt. 分派，分配	8A
alternative	[ɔːl'təːnətiv]	adj. 选择性的，二中择一的	1A
amalgamation	[ə,mælgə'meiʃən]	n. 融合，合并	11B
ambiguity	[,æmbi'gjuːiti]	n. 含糊，不明确	3B
amendment	[ə'mendmənt]	n. 改善，改正	9A
analogy	[ə'nælədʒi]	n. 类似，类推	4B
analyze	['ænəlaiz]	vt. 分析，分解	1A
android	['ændrɔid]	n. 安卓；机器人	5B
antenna	[æn'tenə]	n. 天线	6A
anticipate	[æn'tisipeit]	v. 预订，预见，可以预料	3A
antidote	['æntidəut]	n. 解毒剂，矫正方法	1B
anti-theft	['ænti-θeft]	adj. 防盗的	6A
application	[,æpli'keiʃən]	n. 应用，应用程序	1A
approach	[ə'prəutʃ]	n. 方法，步骤，途径	1A
appropriate	[ə'prəupriət]	adj. 适当的	4A

单词	音标	意义	课次
approved	[əˈpruːvd]	adj. 经核准的，被认可的	7A
architecture	[ˈɑːkitektʃə]	n. 体系结构	1A
assay	[əˈsei]	n.& v. 化验	1B
assemble	[əˈsembl]	vt. 集合，聚集，装配	1B
assign	[əˈsain]	vt. 分配，指派	3B
assistant	[əˈsistənt]	n. 助手，助教 adj. 辅助的，助理的	1A
association	[əˌsəusiˈeiʃən]	n. 协会，联合	1A
assumption	[əˈsʌmpʃən]	n. 假定，设想，担任，承当	9A
asymmetrical	[ˌæsiˈmetrikəl]	adj. 不均匀的，不对称的	9B
attach	[əˈtætʃ]	v. 配属，隶属于	5A
attempt	[əˈtempt]	n. 努力，尝试，企图 vt. 尝试，企图	4B
attraction	[əˈtrækʃən]	n. 吸引，吸引力，吸引人的事物	11A
attractive	[əˈtræktiv]	adj. 吸引人的，有魅力的	10B
attribute	[əˈtribjuː)t]	n. 属性，品质，特征	1A
audio	[ˈɔːdiəu]	adj. 音频的，声频的，声音的	4A
augmented	[ɔːgˈmentid]	adj. 扩张的	1B
authentication	[ɔːˌθentiˈkeiʃən]	n. 认证，证明，鉴定	10A
authorization	[ˌɔːθəraiˈzeiʃən]	n. 授权，认可	10A
automate	[ˈɔːtəmeit]	v. 使自动化，自动操作	5A
autonomic	[ˌɔːtəuˈnɔmik]	adj. 自治的，自律的	3A
autonomous	[ɔːˈtɔnəməs]	adj. 自治的	8A
avail	[əˈveil]	vi. 有益于，有帮助，有用，有利 vt. 有利于 n. 效用，利益	10B
available	[əˈveiləbl]	adj. 可用的，有效的	1A
avatar	[ˌævəˈtɑː]	n. 化身，天神下凡，具体化	1A
awesome	[ˈɔːsəm]	adj. 出色的，非凡的，可怕的	10B
backbone	[ˈbækbəun]	n. 中枢，骨干	2A
backseat	[ˈbæksiːt]	n. 后座，次要位置	5B
badge	[bædʒ]	n. 徽章，证章	6A
band	[bænd]	n. 波段	7A
bandwidth	[ˈbændwidθ]	n. 带宽	7A
barcode	[ˈbɑːkəud]	n. 条形码	1A
bare-metal	[bɛəˈmetl]	n. 裸机	11B
barrier	[ˈbæriə]	n. 障碍；障碍物，栅栏，屏障	1B
baseband	[ˈbeisbænd]	n. 基带	

单 词	音 标	意 义	课次
bat	[bæt]	n. 蝙蝠，球棒	4A
battlefield	['bætlfi:ld]	n. 战场，沙场	8A
beaconing	['bi:kəniŋ]	n. 信标	9A
bear	[bɛə]	v. 负担，忍受，带给	4B
beehive	['bi:haiv]	n. 蜂窝，蜂箱	9B
behavior	[bi'heivjə]	n. 举止，行为	1A
benefit	['benifit]	n. 利益，好处 vt. 有益于，有助于 vi. 受益	1B
bidirectional	[,baidi'rekʃənəl]	adj. 双向的	4A
billboard	['bilbɔ:d]	n. (户外)布告板，广告牌 vt. 宣传	5B
binding	['baindiŋ]	n. 绑定，捆绑	3A
biodegradable	[,baiəudi'greidəbl]	adj. 生物所能分解的	1B
bio-inspired	['baiəu-in'spaiəd]	adj. 仿生的	3A
blackberry	['blækbəri]	n. 黑莓	5B
blimp	[blimp]	n. 软式小型飞船	5B
Bluetooth	['blu:tu:θ]	n. 蓝牙	7B
bluntly	['blʌntli]	adv. 坦率地，率直地	10B
body	['bɔdi]	n. 主体	8A
bottleneck	['bɔtl,nek]	n. 瓶颈	10B
bridge	[bridʒ]	n. 网桥	2B
broadband	['brɔ:dbænd]	n. 宽带	10B
broadcast	['brɔ:dkɑ:st]	n. & v. 广播	7B
broker	['brəukə]	n. 中介，中间设备	11B
brouter	['brautə]	n. 桥式路由器	2B
browse	[brauz]	v. & n. 浏览	5A
browser	[brauzə]	n. 浏览器	10A
bulky	['bʌlki]	adj. 大的，容量大的，体积大的	6B
burden	['bə:dn]	n. 负担	7B
button	['bʌtn]	n. 钮扣，按钮	4A
byte	[bait]	n. 字节	7B
cable	['keibl]	n. 电缆	2A
calculate	['kælkjuleit]	vt. 计算，算出	4A
capability	[,keipə'biliti]	n. (实际)能力，性能，容量，接受力	1A
capacitive	[kə'pæsitiv]	adj. 电容性的	3B

单 词	音 标	意 义	课次
carrier	[ˈkæriə]	n. 载波	
carry	[ˈkæri]	vt. 携带，运送，支持，传送 vi. 被携带，能达到	2A
cashier	[kæˈʃiə]	n. (商店等的)出纳员，收款柜台	1B
cattle	[ˈkætl]	n. 牛，家养牲畜	6B
ceiling	[ˈsiːliŋ]	n. 天花板；最高限度	4A
cellphone	[ˈselfəun]	n. 蜂窝式便携无线电话	5A
centralize	[ˈsentrəlaiz]	vt. 集聚，集中	1A
challenge	[ˈtʃælindʒ]	n. 挑战 vt.向……挑战	11A
chaotic	[keiˈɔtik]	adj. 混乱的，无秩序的，混沌的	1A
characteristic	[ˌkæriktəˈristik]	n. 特性，特征	3A
charge	[tʃɑːdʒ]	n. 负荷，电荷 v. 充电	
checkout	[ˈtʃekaut]	n. 结账台；检验，校验	5A
checksum	[ˈtʃeksʌm]	n. 校验和	5A
chip	[tʃip]	n. 芯片	6A
chunky	[ˈtʃʌŋki]	adj. 粗短的，厚实的	6B
circumstance	[ˈsəːkəmstəns]	n. 环境，详情，境况	6A
circuit-switched	[ˈsəːkit-switʃd]	adj. 线路交换的	2B
civilian	[siˈviljən]	adj. 民用的，平民的，民间的 n. 平民	8B
classified	[ˈklæsifaid]	v. 分类 adj. 机密的	5A
client	[ˈklaiənt]	n. 顾客，客户，客户机	2A
closure	[ˈkləuʒə]	n. 关闭 vt. 使终止	8B
co-design	[ˈkəu-diˈzain]	v. 协作设计，合作计划	11A
coexist	[ˌkəuigˈzist]	vi. 共存	1A
coexistence	[ˌkəuigˈzistəns]	n. 共存	1B
coherent	[kəuˈhiərənt]	adj. 一致的，连贯的	3A
coil	[kɔil]	v. 盘绕，卷	6B
collaboration	[kəˌlæbəˈreiʃən]	n. 协作	11A
collaborative	[kəˈlæbərətiv]	adj. 合作的，协作的，协力完成的	11A
collapse	[kəˈlæps]	n. 倒塌，崩溃，失败 vi. 倒塌，崩溃，瓦解，失败	8A
colleague	[ˈkɔliːg]	n. 同事，同僚	4B
collective	[kəˈlektiv]	adj. 集体的 n. 集体	11A
collision	[kəˈliʒən]	n. 碰撞，冲突	6A
colocation	[ˌkəuləuˈkeiʃən]	n. 场地出租	11B
commercial	[kəˈməʃəl]	adj. 商业的，商用的	8B

单词	音标	意义	课次
communication	[kə,mju:ni'keiʃn]	n. 通信	1A
community	[kə'mju:niti]	n. 公社，团体，社会	11A
commute	[kə'mju:t]	v. 交换，抵偿	4B
compactor	[kəm'pæktə]	n. 垃圾捣碎机，压土机，夯土机	11A
compatibility	[kəm,pæti'biliti]	n. 兼容性	
compatible	[kəm'pætəbl]	adj. 兼容的	5A
complaint	[kəm'pleint]	n. 投诉；诉苦，抱怨，牢骚	6B
complex	['kɔmpleks]	adj. 复杂的，合成的，综合的 n. 联合体	5B
compliance	[kəm'plaiəns]	n. 依从，顺从	11B
complicate	['kɔmplikeit]	v. (使)变复杂	3B
component	[kəm'pəunənt]	n. 成分 adj. 组成的，构成的	2A
composite	['kɔmpəzit]	adj. 合成的，复合的 n. 合成物	3B
compromise	['kɔmprəmaiz]	v. 危及……的安全	10A
conceive	[kən'si:v]	vt. 构思，以为，持有 vi. 考虑，设想	9A
concentration	[,kɔnsen'treiʃən]	n. 浓度	8B
concrete	['kɔnkri:t]	adj. 具体的，有形的	3A
confidence	['kɔnfidəns]	n. 信心	5A
configurable	[kən'figərəbl]	adj. 结构的，可配置的	3A
configuration	[kən,figju'reiʃən]	n. 构造，结构，配置	3B
configure	[kən'figə]	vi. 配置，设定	1A
confined	[kən'faind]	adj. 被限制的，狭窄的	7B
conform	[kən'fɔ:m]	vt. 使一致，使遵守，使顺从 vi. 符合，相似，适应环境 adj. 一致的，顺从的	9A
congestion	[kən'dʒestʃən]	n. 拥塞，拥挤	
connect	[kə'nekt]	v. 连接，联合，关连	3A
connection	[kə'nekʃən]	n. 连接，关系，接线，线路	2A
connectivity	[kənek'tiviti]	n. 连通性	10B
consequently	['kɔnsikwəntli]	adv. 从而，因此	1A
considerably	[kən'sidərəbli]	adv. 相当地	7A
consideration	[kənsidə'reiʃən]	n. 考虑，需要考虑的事项	11B
consistency	[kən'sistənsi]	n. 一致性，连贯性	3B
console	[kən'səul]	n. 控制台	7B
conspicuous	[kən'spikjuəs]	adj. 显著的	10B
constant	['kɔnstənt]	adj. 不变的，持续的，坚决的	11A
constraint	[kən'streint]	n. 约束，强制	1A

单　词	音　标	意　义	课次
construct	[kənˈstrʌkt]	vt. 建造，构造，创立	3B
consult	[kənˈsʌlt]	v. 翻阅，查阅，参考，考虑	1B
consumption	[kənˈsʌmpʃən]	n. 消费，消费量	11A
container	[kənˈteinə]	n. 容器，集装箱	6A
contention	[kənˈtenʃən]	n. 争用，争夺	9A
continuous	[kənˈtinjuəs]	adj. 连续的，持续的	11B
continuously	[kənˈtinjuəsli]	adv. 不断地，连续地	9B
controllable	[kənˈtrəuləbl]	adj. 可管理的，可操纵的，可控制的	3A
controversy	[ˈkɔntrəvə:si]	n. 论争，辩论，论战	11A
conventional	[kənˈvenʃənl]	adj. 惯例的，常规的，习俗的，传统的	4A
convergence	[kənˈvə:dʒəns]	n. 汇聚，集中，集合	9A
conversation	[ˌkɔnvəˈseiʃən]	n. 会话，交谈	2A
converse	[kənˈvə:s]	vi. 认识；谈话，交谈	1A
convert	[kənˈvə:t]	vt. 使转变，转换……	2A
converter	[kənˈvɜ:tə]	n. 变频器	
cooperative	[kəuˈɔpərətiv]	adj. 合作的，协力的	1A
coordinate	[kəuˈɔ:dinit]	n. 同等者，同等物，坐标(用复数) adj. 同等的，并列的 vt. 调整，整理	4A
coordinator	[kəuˈɔ:dineitə]	n. 协调者	9A
corporation	[ˌkɔ:pəˈreiʃən]	n. 公司；企业	10A
correlation	[ˌkɔriˈleiʃən]	n. 相关(性)	11B
corresponding	[ˌkɔrisˈpɔndiŋ]	adj. 相应的	8A
cost	[kɔst]	n. 成本，价钱，代价	2B
counterargument	[ˈkauntəˌɑ:gjəmənt]	n. 辩论，抗辩	10A
countermeasure	[ˈkauntəˌmeʒə]	n. 对策，反措施	3A
counterpart	[ˈkauntəpɑ:t]	n. 副本，极相似的人或物，配对物	5B
countryside	[ˈkʌntrisaid]	n. 乡下地方，乡下居民	11A
coupler	[ˈkʌplə]	n. 耦合器	
coverage	[ˈkʌvəridʒ]	n. 覆盖	7A
coworker	[ˈkəuˌwə:kə]	n. 合作者，同事	4A
credential	[kriˈdenʃəl]	n. 信任状	4B
critical	[ˈkritikəl]	adj. 极重要的，至关紧要的	1A
crosstalk	[ˈkrɔsˌtɔ:k]	n. 串话，串扰	
custom-built	[ˈkʌstəmˈbilt]	adj. 定制的，定做的	6A

单　词	音　标	意　义	课次
customer	['kʌstəmə]	n. 消费者	4B
cyberobject	['saibəɔb'dʒikt]	n. 计算体，计算部件	1A
cyberville	['saibəvil]	n. 网络城市	11A
dangerous	['deindʒərəs]	adj. 危险的	5A
data	['deitə]	n. 数据	2A
database	['deitəbeis]	n. 数据库，资料库	5A
dealer	['di:lə]	n. 经销商，商人	6A
debugging	[di:'bʌgiŋ]	n. 调试	8A
decade	['dekeid]	n. 十年，十	5A
decentralize	[di:'sentrəlaiz]	n. 分散	3A
decidedly	[di'saididli]	adv. 果断地，断然地	9B
decipherment	[di'saifəmənt]	n. 解密	
decode	[,di:'kəud]	vt. 解码，译解	7A
dedicated	['dedikeitid]	adj. 专注的	4B
deduct	[di'dʌkt]	vt. 扣除	6B
definitely	['definitli]	adv. 明确地，干脆地	6B
definition	[,defi'niʃən]	n. 定义，解说，精确度	1A,7B
delegate	['deligit]	n. 代表　vt. 委派……为代表	1B
delegation	[,deli'geiʃən]	n. 授权，委托	3B
dense	[dens]	adj. 密集的	3B
deny	[di'nai]	v. 否认，拒绝	10A
departure	[di'pɑ:tʃə]	n. 启程，出发，离开	5A
dependable	[di'pendəbl]	adj. 可靠的	4B
deploy	[di'plɔi]	v. 展开，配置	1A,8B
describe	[dis'kraib]	v. 描述	7B
destination	[,desti'neiʃən]	n. 目的地；[计]目的文件，目的单元格	5A
detailed	['di:teild]	adj. 详细的，逐条的	5A
detect	[di'tekt]	vt. 探测，发觉	7A
detection	[di'tekʃən]	n. 察觉，发觉，探测，发现	8B
device	[di'vais]	装置，设备	2A
devise	[di'vaiz]	vt. 设计，发明，图谋，做出(计划)，想出(办法)	5B
digital	['didʒitl]	adj. 数字的，数位的　n. 数字，数字式	10A
dimension	[di'menʃən]	n. 维(数)，度(数)，元	11A
direction	[di'rekʃən]	n. 收件人地址	2B

单词	音标	意义	课次
directionality	[diˌrekʃən'næliti]	n. 方向性，定向性	8A
disadvantage	[ˌdisəd'vɑ:ntidʒ]	n. 不利条件，缺点，劣势	6B
disconnected	[ˌdiskə'nektid]	adj. 分离的，离散的，不连贯的	3A
disconnecting	[ˌdiskə'nektiŋ]	n. 拆开，解脱，分离	7A
discount	['diskaunt]	n. 折扣	1B
discriminatory	[di'skriminətəri]	adj. 有差别的	3B
discussion	[dis'kʌʃen]	n. 讨论	2A
disease	[di'zi:z]	n. 疾病，弊病	1B
displace	[dis'pleis]	vt. 取代，置换	3B
disposable	[dis'pəuzəbl]	adj. 可任意使用的	3A
dispute	[dis'pju:t]	v. 争论，辩论，怀疑，阻止 n. 争论，辩论，争吵	1B
disruption	[dis'rʌpʃen]	n. 中断，分裂，瓦解，破坏	3A
distil	[di'stil]	vt. 提取，精炼，浓缩；融入	11A
distorted	[dis'tɔ:tid]	adj. 扭歪的，受到曲解的	5B
distributed	[dis'tribju:tid]	adj. 分布式的	3A
diversity	[dai'və:siti]	n. 差异，多样性	1B
dominion	[də'minjən]	n. 域	3B
dosage	['dəusidʒ]	n. 剂量，配药，用量	1B
doubtless	['dautlis]	adj. 无疑的，确定的 adv. 无疑地，确定地	10B
downlink	['daunliŋk]	n. 下行线	10B
download	['daunləud]	v. 下载	10B
draft	[drɑ:ft]	n. 草稿，草案，草图　v. 草拟	7A
dramatically	[drə'mætikəli]	adv. 戏剧地，引人注目地	10B
drug	[drʌg]	n. 药	1B
duplexer	['dju:pleksə]	n. 双工器	
duplicate	['dju:plikeit]	adj. 复制的	5A
dust	[dʌst]	n. 灰尘	1B
dynamic	[dai'næmik]	adj. 动态的	1A
ecosystem	[i:kə'sistəm]	n. 生态系统	1A
edible	['edibl]	adj. 可食用的	5B
efficiently	[i'fiʃəntli]	adv. 有效率地，有效地	4B
e-governance	[i:-'gʌvənəns]	n. 电子治理，电子管理，电子政务	10B
electromagnetic	[ilektrəu'mægnitik]	adj. 电磁的	6B

单词	音标	意义	课次
element	['elimənt]	n. 要素，元素，成份，元件	2A
eliminate	[i'limineit]	vt. 排除，消除	7A
embankment	[im'bæŋkmənt]	n. 堤防，筑堤	8B
embedded	[em'bedid]	adj. 植入的，嵌入的，内含的	1A
embedding	[em'bediŋ]	n. 嵌入，埋入	11A
emergent	[i'mə:dʒənt]	adj. 紧急的，突然出现的，自然发生的	8A
emission	[i'miʃən]	n. 发射	
emphasis	['emfəsis]	n. 强调，重点	9A
employee	[,emplɔ'i:]	n. 雇工，雇员，职工	10A
encapsulate	[in'kæpsjuleit]	v. 封装	4B
encapsulation	[in,kæpsju'leiʃən]	n. 封装，包装	4B
encase	[in'keis]	vt. 包括，装入，包住，围	6A
encoder	[in'kəudə]	n. 译码器，编码器	
encrypt	[in'kript]	v. 加密，将……译成密码	4B
end-user	['end,ju:zə]	n. 终端用户	8A
energy	['enədʒi]	n. 精力，活力；能量	6A
engage	[in'geidʒ]	vt. 吸引	6B
ensure	[in'ʃuə]	v. 确保，保证	6A
enterprise	['entəpraiz]	n. 企业；公司	2B
entertainment	[entə'teinmənt]	n. 娱乐	9B
enthusiastic	[in,θju:zi'æstik]	adj. 热心的，热情的	5A
entity	['entiti]	n. 实体	3B
environmental	[in,vaiərən'mentl]	adj. 周围的，环境的	1A
envision	[in'viʒən]	vt. 想象，预想	10B
e-participation	[i:-pɑ:,tisi'peiʃən]	n. 电子分享，电子参与	11A
equally	['i:kwəli]	adv. 相等地，平等地，公平地	4B
error	['erə]	n. 误差	5A
essentially	[i'senʃəli]	adv. 本质上，本来	4B
establish	[is'tæbliʃ]	vt. 建立，设立，确定	2A
Ethernet	['i:θənet]	n. 以太网	2B
eventually	[i'ventʃuəli]	adv. 终于，最后	2A
evidence	['evidəns]	n. 明显，显著，明白，迹象	5A
evolve	[i'vɔlv]	v. (使)发展，(使)进展，(使)进化	1A
examine	[ig'zæmin]	vt. 检查，调查；研究，分析	2B
exceed	[ik'si:d]	vt. 超越，胜过 vi. 超过其他	11B

单　　词	音　　标	意　　义	课次
exception	[ik'sepʃən]	n. 除外，例外，反对，异议	1A
exchange	[iks'tʃeindʒ]	vt. & n. 交换，调换，兑换，交易	1A
executive	[ig'zækjutiv]	n. 总经理；行政部门；[计算机]执行指令 adj. 执行的；管理的；政府部门的	10A
exit	['eksit]	n. 出口，退场　vi. 退出，脱离	6B
expand	[iks'pænd]	v. 扩展	4B
expensive	[iks'pensiv]	adj. 花费的，昂贵的	5B
experience	[iks'piəriəns]	n. & vt. 经验，体验，经历	10A
experimental	[eks,peri'mentl]	adj. 实验的，根据实验的	11A
explode	[iks'pləud]	vi. 爆发，破除，推翻，激发	5B
exploit	[iks'plɔit]	v. 使用	9A
exploitation	[,eksplɔi'teiʃən]	n. 使用	1A
extensive	[iks'tensiv]	adj. 广大的，广阔的，广泛的	9B
extranet	['ekstrənet]	n. 外联网	1A
facilitate	[fə'siliteit]	vt. 使便利，帮助，使容易，促进	1A
fade	[feid]	vi. 减弱下去，消失	6A
fading	['feidiŋ]	n. 衰退，衰落	
fairground	['fɛəgraund]	n. 举行赛会的场所，露天市场	5A
favor	['feivə]	vt. 支持，赞成，促成	9B
federate	['fedərit]	adj. 同盟的，联合的	3A
feedback	['fi:dbæk]	n. 反馈，反应	1B
ferry	['feri]	n. 摆渡，渡船，渡口 vt. 渡运，(乘渡船)渡过，运送 vi. 摆渡	4B
fiber	['faibə]	n. 光纤	2A
file	[fail]	n. 文件	2A
filter	['filtə]	n. 滤波器，过滤器，滤光器；筛选 vt. 过滤，渗透	7A
finder	['faində]	adj. 发现者，探测器	5B
fingerprint	['fiŋgəprint]	n. 指纹，手印　vt. 采指纹	6B
firefighter	['faiəfaitə]	n. 消防队员	8B
firefighting	['faiəfaitiŋ]	n. 消防	2B
fleet	[fli:t]	n. 舰队，港湾　adj. 快速的，敏捷的	4B
flexibility	[,fleksə'biliti]	n. 弹性，适应性，机动性，柔性	1B
flexible	['fleksəbl]	adj. 柔韧性，灵活的，柔软的	2A
flexicity	[fleksi'siti]	n. 弹性城市，柔性城市	11A

单 词	音 标	意 义	课次
flocking	['flɔkiŋ]	n. 群集，成群结队而行	8A
flood	[flʌd]	n. 洪水，水灾 vt. 淹没 vi. 被水淹，涌进	2B
fluctuation	[,flʌktju'eiʃən]	n. 波动，起伏	5A
flyover	['flaiəuvə]	n. 天桥，立交桥	8B
foolproof	['fu:lpru:f]	adj. 十分简单的，十分安全的，极坚固的	7A
foreseeable	[fɔ:'si:əbl]	adj. 可预知的，能预测的	2A
format	['fɔ:mæt]	n. 形式，格式 vt. 安排……的格局(或规格)，[计]格式化（磁盘）	5B
forward	['fɔ:wəd]	vt. 转发，传输	8A
foster	['fɔstə]	vt. 培养，鼓励	3A
fragment	['frægmənt]	n. 碎片，断片，片段	3B
frame	[freim]	n. 帧，框架，画面	2B
framework	['freimwə:k]	n. 构架，框架，结构	1A
fraudulent	['frɔ:djulənt]	adj. 欺诈的，欺骗性的	5A
frequency	['fri:kwənsi]	n. 频率，周率	7A
frontrunner	['frʌntrʌnə]	n. 领跑者，前锋	6B
functional	['fʌŋkʃənl]	adj. 功能的	1A
functionality	[,fʌŋkʃə'næliti]	n. 功能性，泛函性	9B
fundamental	[,fʌndə'mentl]	adj 基础的，基本的 n. 基本原则，基本原理	9A
fundamentally	[fʌndə'mentəli]	adv. 基础地，根本地	1B
furnish	['fə:niʃ]	v. 供给	2B
garbage	['gɑ:bidʒ]	n. 垃圾，废物	11A
gateway	['geitwei]	n. 门，通路，网关	8A
gear	[giə]	v. 调整，(使)适合 n. 装置	7A
generality	[,dʒenə'ræliti]	n. 一般性，普遍性，大部分	9A
generate	['dʒenə,reit]	vt. 产生，发生	11B
generator	['dʒenəreitə]	n. 发生器	5B
generic	[dʒi'nerik]	adj. 一般的，普通的	9B
geo-fencing	[dʒi:əu-'fensiŋ]	n. 地理围栏	8B
geographic	[dʒiə'græfik]	adj. 地理学的，地理的	1A
geospatial	['dʒi:əuspeiʃəl]	adj. 地理空间的	1A
gigabit	['dʒigəbit]	n. 吉比特	10B
given	['givn]	adj. 约定的；特定的；指定的	2B

单 词	音 标	意 义	课次
glacier	['glæsiə]	n. 冰山	8B
granular	['grænjulə]	adj. 由小粒而成的，粒状的	3A
greenhouse	['gri:haus]	n. 温室	8B
grocery	['grəusəri]	n. <美>食品杂货店，食品，杂货	6B
guarantee	[,gærən'ti:]	vt. 保证，担保 n. 保证，保证书，担保，抵押品	9A
guess	[ges]	v. 猜测，推测，想，认为 n. 猜测，推测	5B
gum	[gʌm]	n. 口香糖，香口胶，泡泡糖	5B
hacker	['hækə]	n. 电脑黑客	7A
handheld	['hænd,held]	adj. 掌上型，手持型	6A
handle	['hændl]	vt. 处理，操作 n. 句柄	10A
handling	['hændliŋ]	n. 处理 adj. 操作的	4B
handy	['hændi]	adj. 手边的，就近的，便利的，敏捷的，容易取得的	4A
hardware	['hɑ:dwɛə]	n. (电脑的)硬件	2A
harness	['hɑ:nis]	n. 系在身上的绳子 vt. 上马具，披上甲胄	9A
harsh	[hɑ:ʃ]	adj. 粗糙的，荒芜的，苛刻的	8A
hazardous	['hæzədəs]	adj. 危险的，冒险的	8B
header	['hedə]	n. 报头	
headset	['hedset]	n. 戴在头上的耳机或听筒，头套	7B
healing	['hi:liŋ]	n. 康复，复原 adj. 有治疗功用的	3A
healthcare	['helθkɛə]	n. 医疗保健，健康护理，卫生保健	5A
heighten	['haitn]	v. 提高，升高	11B
hesitate	['heziteit]	v. 犹豫，踌躇，不愿	10A
heterogeneity	[,hetərəudʒi'ni:iti]	n. 异种，异质，不同成分	8A
heterogeneous	[,hetərəu'dʒi:niəs]	adj. 不同种类的，异类的	10A
heterogeneous	[,hetərəu'dʒi:niəs]	adj. 不同种类的，异类的	11B
hexagon	['heksəgən]	n. 六角形，六边形	5A
highlight	['hailait]	vt. 突出	1B
highway	['haiwei]	n. 公路，大路，高速公路	4B
hinder	['hində]	adj. 后面的 v. 阻碍，打扰	5B

单词	音标	意义	课次
hire	[ˈhaiə]	n. 租金；酬金，工钱；[非正式用语] 被雇佣的人 vt. 聘用；录用；雇用；租用 vi. 受雇；得到工作	10A
hobby	[ˈhɔbi]	n. 业余爱好	11A
holder	[ˈhəuldə]	n. 持有者，占有者，(台、架等)支持物，固定器	5A
holistic	[həuˈlistik]	adj. 整体的，全盘的	11A
hop	[hɔp]	v. 中继段	2B
host	[həust]	n. 主机 v. 做主机	1A
hostile	[ˈhɔstail]	adj. 敌对的，敌方的	8A
hotspot	[ˈhɔtspɔt]	n. 热点，热区	7A
household	[ˈhaushəuld]	n. 家庭，家族 adj. 家庭的，家族的，普通的，平常的	1B
hub	[hʌb]	n. 网络集线器，网络中心	7A
huge	[hju:dʒ]	adj. 巨大的，极大的，无限的	11B
humidity	[hju:ˈmiditi]	n. 湿度	8B
hurdle	[ˈhə:dl]	n. 障碍 v. 克服(障碍)	5B
hybrid	[ˈhaibrid]	n. 混合物 adj. 混合的	11B
hyperlink	[ˈhaipəliŋk]	n. 超链接	5A
hypertext	[ˈhaipətekst]	n. 超文本	2A
identifiable	[aiˈdentifaiəbl]	adj. 可以确认的	1A
identification	[ai,dentifiˈkeiʃən]	n. 辨认，鉴定，证明	3B
identifier	[aiˈdentifaiə]	n. 标识符	3B
identify	[aiˈdentifai]	vt. 识别，鉴别	1A
identity	[aiˈdentiti]	n. 身份，特性	1A
ignore	[igˈnɔ:]	vt. 不理睬，忽视	5A
image	[ˈimidʒ]	n. 图像	5A
imager	[ˈimidʒə]	n. 成像器	5A
imagine	[iˈmædʒin]	vt. 想象，设想	6B
imminent	[ˈiminənt]	adj. 即将来临的，逼近的	10B
immunity	[iˈmju:niti]	n. 抗扰度	
impact	[ˈimpækt]	n. 冲击，影响 vt. 对……发生影响	11A
implementation	[,implimenˈteiʃən]	n. 执行	9A
importance	[imˈpɔ:təns]	n. 重要(性)，重大，价值	10B
improvement	[imˈpru:vmənt]	n. 改进，进步	1B

单 词	音 标	意 义	课次
impulse	['impʌls]	n. 脉动	
inaccessible	[inæk'sesibl]	adj. 不能进入的，不能达到的	8B
inclusion	[in'klu:ʒən]	n. 包含，内含物	8A
incompatible	[,inkəm'pætəbl]	adj. 不兼容的；不相容的；互斥的	1B
independent	[indi'pendənt]	adj. 独立的	1A
index	['indeks]	n. 索引 vi. 做索引	1A
induct	[in'dʌkt]	v. 感应	6B
inefficiency	[,ini'fiʃənsi]	n. 无效率，无能	11A
inductive	[in'dʌktiv]	adj. 电感的，感应的	3B
inductively	[in'dʌktivli]	adv. 感应地，诱导地，归纳地	6B
inevitable	[in'evitəbl]	adj. 不可避免的，必然的	10B
influence	['influəns]	n. 影响，(电磁)感应 vt. 影响，改变	1A
influx	['inflʌks]	n. 流入	11B
infrared	['infrə'red]	adj. 红外线的 n. 红外线	7B
infrastructure	['infrəstrʌktʃə]	n. 基础，下部构造，基础组织	1A
infringe	[in'frindʒ]	v. 破坏，侵犯，违反	1B
inherent	[in'hiərənt]	adj. 固有的，内在的，与生俱来的	8A
initiative	[i'niʃiətiv]	n. 倡议；主动性；主动精神 adj.自发的；创始的；初步的	11A
innovation	[,inəu'veiʃən]	n. 改革，创新	1B
insert	[in'sə:t]	vt. 插入，嵌入 n. 插入物	6A
insight	['insait]	n 洞察力，见识	11B
inspect	[in'spekt]	vt. 检查，审查	2B
inspection	[in'spekʃən]	n. 视察	8B
install	[in'stɔ:l]	vt. 安装，安置	8B
instance	['instəns]	n. 实例，建议，要求，情况，场合	3A
instantaneous	[,instən'teinjəs]	adj. 瞬间的，即刻的，即时的	5B,7B
instrument	['instrəmənt]	n. 工具，器械，器具，手段	11B
instrumentation	[,instrumen'teiʃən]	n. 使用仪器	11A
insurance	[in'ʃuərəns]	n. 保险，保险单，保险业，保险费	1B
integrate	['intigreit]	vt. 使成整体，使一体化	1A
interaction	[,intər'ækʃən]	n. 交互作用，交感	1A
interactive	[,intər'æktiv]	adj. 交互式的	10B
interactivity	[intə'æktiviti]	n. 交互，互动；人机对话	5B
intercept	[,intə'sept]	vt. 中途阻止，截取	4B

单 词	音 标	意 义	课次
interception	[ˌintə(:)'sepʃən]	n. 监听，拦截，侦听	
intercharacter	[in'tə:'kæriktə]	adj. 字符间的	5A
intercom	['intəkɔm]	n. 联络所用对讲电话装置，内部通信联络系统	7B
interconnect	[ˌintəkə'nekt]	vt. 使互相连接	2B
interface	['intəˌfeis]	n. 接口，界面	10A
interfacing	['intəfeisiŋ]	n. 界面连接，接口连接	8A
interference	[ˌintə'fiərəns]	n. 冲突，干涉	7A
interior	[in'tiəriə]	adj. 内部的，内的 n. 内部	5B
interleaving	[ˌintə(:)'li:viŋ]	n. 交叉，交错	5A
intermediate	[ˌintə'mi:djət]	adj. 中间的 n. 媒介	11B
intermittent	[ˌintə(:)'mitənt]	adj. 间歇的，断断续续的	9B
intermodulation	[ˌintə(:)ˌmɔdju'leiʃən]	adj. 互调	
internal	[in'tə:nl]	adj. 内在的，内部的	4B
Internet	['intənet]	n. 因特网，国际互联网络	2A
interoperability	['intərˌɔpərə'biləti]	n. 互用性，协同工作的能力	3A
interoperable	[ˌintər'ɔpərəbl]	adj. 能共同操作的，能共同使用的	1A
interpretation	[inˌtə:pri'teiʃən]	n. 解释，阐明	11A
interpretive	[in'tə:pritiv]	adj. 作为说明的，解释的	5A
interval	['intəvəl]	n. 时间间隔	7B
intervention	[ˌintə(:)'venʃən]	n. 干涉	1B
interwork	[ˌintə'wə:k]	v. 互连	
intranet	['intrənet]	n. 内联网	1A
introduction	[ˌintrə'dʌkʃən]	n. 介绍，传入	5A
intruder	[in'tru:də]	n. 入侵者	9B
invasion	[in'veiʒən]	n. 入侵	6A
inventory	['invəntri]	vt. 编制……的目录；盘存，清查 vi. 对清单上存货的估价 n. 详细目录，存货，财产清册，总量	1A
issue	['isju:]	n. 问题，结果	1A
item	['aitem]	n. 项目	5A
jeopardise	['dʒepədaiz]	v. 使受危险，危及	3B
jumble	['dʒʌmbl]	v. 混杂，搞乱 n. 混乱	5B
jurisdiction	[ˌdʒuəris'dikʃən]	n. 管辖权，权限	9B
kernel	['kə:nl]	n. 核	8A
keying	['ki:iŋ]	n. 键控	

单词	音标	意义	课次
kick-start	[kik-stɑːt]	vt. 发起　n. 强劲推动力	5B
knowledgeable	['nɔlidʒəbl]	adj. 知识渊博的，有见识的	4B
knowledge-intensive	['nɔlidʒin'tensiv]	adj. 知识密集型的	11A
label	['leibl]	n. 标签，签条，商标，标志 vt. 贴标签于，指……为，分类，标注	5A
landslide	['lændslaid]	n. 山崩，山体滑坡	8B
latency	['leitənsi]	n. 反应时间	9B
layer	['leiə]	n. 层	3A
leakage	['liːkidʒ]	n. 漏，泄漏，渗漏	3A
leverage	['liːvəridʒ]	n. 手段；杠杆作用	10A
license	['laisəns]	n. 许可(证)，执照　v. 许可	10A
limited	['limitid]	adj. 有限的	5A
limitless	['limitlis]	adj. 无限的，无界限的	10A
linear	['liniə]	adj. 线的，直线的，线性的	5A
liquid	['likwid]	n. 液体，流体	1B
location	[ləu'keiʃən]	n. 位置，场所	1A
logging	['lɔgiŋ]	n. 存入，联机；记录	2B
logistics	[lə'dʒistiks]	n. 物流	5A
logo	['lɔgəu]	n. 标识	5B
logon	['ləugən, 'lɔgɔn]	v 登录上网	2B
longevity	[lɔn'dʒeviti]	n. 寿命	3B
luggage	['lʌgidʒ]	n. 行李，皮箱	5A
luthier	['luːtiə]	n. 拨弦乐器制作匠	6A
maintain	[men'tein]	vt. 维持，维修	4B
malicious	[mə'liʃəs]	adj. 怀恶意的，恶毒的	3A
malleable	['mæliəbl]	adj. 有延展性的	2A
mandate	['mændeit]	n. 命令，要求	7B
massive	['mæsiv]	adj. 可观的，巨大的，大量的	1A
match	[mætʃ]	n. 匹配	2A
matrix	['meitriks]	n. 矩阵	5A
maximization	[,mæksəmai'zeiʃən]	n. 最大值化，极大值化	8A
maximum	['mæksiməm]	n. 最大量，最大限度，极大 adj. 最高的，最多的，最大极限的	8A
meaningful	['miːniŋful]	adj. 有意义的	2A

单词	音标	意义	课次
measure	['meʒə]	n. 量度器，量度标准，方法，测量，措施 vt. 测量，测度，估量	4A
mechanism	['mekənizəm]	n. 机械装置，机构，机制	1B
mediation	[,mi:di'eiʃən]	n. 仲裁，调停，调解	1A
medication	[,medi'keiʃən]	n. 药物治疗，药物处理，药物	5A
medium	['mi:djəm]	n. 媒体，方法，媒介 adj. 中间的，中等的	5B
memory	['meməri]	n. 存储器，内存	3A
merchandise	['mə:tʃəndaiz]	n. 商品，货物	6A
merchandising	['mə:tʃəndaiziŋ]	n. 商品之广告推销，销售规划	5A
metadata	[metə'deitə]	n. 元数据	8A
metal	['metl]	n. 金属	1B
microchip	['maikrəutʃip]	n. 微芯片	6A
microcontroller	[,maikrəukən'trəulə]	n. 微控制器	8A
microprocessor	[maikrəu'prəusesə]	n. 微处理机	
microscopic	[maikrə'skɔpik]	adj. 用显微镜可见的，精微的	8A
microwave	['maikrəuweiv]	n. 微波	
middleware	['midlwεə]	n. 中间设备，中间件	10A
military	['militəri]	adj. 军事的，军方的，军队的	8B
millimeter	['milimi:tə]	n. 毫米	6A
miniaturization	[,miniətʃərai'zeiʃən]	n. 小型化	3A
minimize	['minimaiz]	vt. 将……减到最少 v. 最小化	10B
minimum	['minimam]	adj. 最小的，最低的 n. 最小值，最小化	8A
minuscule	[mi'nʌskju:l]	adj. 极小的	1A
misland	[mis'lænd]	vt. 放错，遗失	1A
misuse	[mis'ju:z] [mis'ju:s]	v. 误用，错用，滥用 n. 误用，错用；滥用	1B
mobility	[məu'biliti]	n. 活动性，灵活性，机动性	10B
modality	[məu'dæliti]	n. 形式，形态，特征	3B
mode	[məud]	n. 模式，方式，样式	7B
modem	['məudəm]	n. 调制解调器	8A
modest	['mɔdist]	adj. 适度的	9A
modulation	[,mɔdju'leiʃən]	n. 调制	7A
modulator	['mɔdjuleitə]	n. 调节器，调制器	

单 词	音 标	意 义	课次
module	[ˈmɔdjuːl]	n. 模块	3A
monitor	[ˈmɔnitə]	vt. 监控 n. 监视器，监控器	3A
monolithic	[ˌmɔnəˈliθik]	n. 单片电路，单块集成电路	3A
mote	[məut]	n. 尘埃，微粒	8A
motivate	[ˈməutiveit]	v. 推动，激发	8A
multiframe	[ˈmʌltifreim]	n. 多帧	
multihop	[ˈmʌltihɔp]	n. 多次反射	9A
multipath	[ˈmʌltipɑːθ]	n. 多径	
multiple	[ˈmʌltipl]	adj 多样的，多重的 n. 倍数，若干 v. 成倍增加	1B
multiplexing	[ˈmʌltipleksiŋ]	n. 多路复接	
multithreading	[ˈmʌltiˈθrediŋ]	n. 多线程，多线索	8A
multitier	[ˈmʌltitiə]	n. 多层，多列	3B
nano-electronics	[ˈnænəu-ilekˈtrɔniks]	n. 纳电子学	3A
nerve	[nəːv]	n. 神经	11A
neutral	[ˈnjuːtrəl]	adj. 中立的	3A
newsletter	[ˈnjuːzˌletə]	n. 时事通讯	6B
Nintendo	[ninˈtendəu]	n. 任天堂(游戏名)	7B
node	[nəud]	n. 节点	2B
nondeterministic	[nɔnditəːmiˈnistik]	adj. 非定常的，非确定的	1A
nonexclusive	[ˈnɔniksˈkluːsiv]	adj. 非独家的	3B
non-repudiable	[nɔŋ-riˈpjuːdiəbl]	adj. 不可否认的	11B
non-toxic	[nʌnˈtɔksik]	adj. 无毒的	3A
notation	[nəuˈteiʃən]	n. 符号	8A
notify	[ˈnəutifai]	v. 通报	8B
occasional	[əˈkeiʒnəl]	adj. 偶然的，非经常的，特殊场合的，临时的	3B
occasionally	[əˈkeiʒənəli]	adv. 有时候，偶尔	10A
occurrence	[əˈkərəns]	n. 出现，发生	8B
off-limit	[ˈɔːf-ˈlimit]	adj. 禁止进入的	5B
online	[ˈɔnlain]	n. 联机，在线式	11A
opaque	[əuˈpeik]	n. 不透明物 adj. 不透明的	3B
opportunity	[ˌɔpəˈtjuːniti]	n. 机会，时机	3A
optical	[ˈɔptikəl]	adj. 光学的	5A
optimize	[ˈɔptimaiz]	vt. 使最优化	1B
option	[ˈɔpʃən]	n. 选项，选择	8A

单 词	音 标	意 义	课次
organ	[ˈɔːgən]	n. 元件，机构	11A
organization	[ˌɔːgənaiˈzeiʃən]	n. 组织，机构，团体	5A
original	[əˈridʒənəl]	adj. 最初的，原始的，独创的，新颖的	1A
originally	[əˈridʒənəli]	adv. 最初，原先	5A
oscillate	[ˈɔsileit]	v. 振荡	1B
outbreak	[ˈautbreik]	n. (战争的)爆发，(疾病的)发作	1B
outgoing	[ˈautgəuiŋ]	n. 外出，开支，流出	4B
outlet	[ˈautlet]	n. 出口，出路	6B
outsource	[ˈautsɔːs]	vt. 把……外包 vi. 外包	4B
overdose	[ˈəuvədəus]	n. 配药量过多 vt. 配药过量，使服药过量	1B
overlap	[ˈəuvəˈlæp]	v. (与……)交叠	4A
overlapping	[ˈəuvəˈlæpiŋ]	n. 重叠，搭接	1A
overproduction	[ˈəuvəprəˈdʌkʃən]	n. 生产过剩	1B
ownership	[ˈəunəʃip]	n. 所有权，物主身份	4A
packet	[ˈpækit]	n. 数据包	2A
paradigm	[ˈpærədaim]	n. 范例	8A
paradigmatic	[ˌpærədigˈmætik]	adj. 范式的，例证的	10B
parallel	[ˈpærəlel]	adj. 平行的，相同的，类似的，并联的 n. 平行线，类似，相似物 v. 相应，平行	5A
parameter	[pəˈræmitə]	n. 参数，参量	7B
parameterize	[pəˈræmitəraiz]	vt. 确定……的参数，用参数表示	7B
parcel	[ˈpɑːsl]	n. 小包，包裹 vt. 打包，捆扎	5A
participant	[pɑːˈtisipənt]	n. 参与者，共享者	1A
participation	[pɑːˌtisiˈpeiʃən]	n. 分享，参与	11A
participatory	[pɑːˈtisipeitəri]	adj. 供人分享的	11A
particular	[pəˈtikjulə]	n. 细节，详细 adj. 特殊的，特别的，详细的，精确的	4A
partner	[ˈpɑːtnə]	n. 合伙人，股东，伙伴 vt. 与……合伙，组成一对 vi. 做伙伴，当助手	4B
password	[ˈpɑːswəːd]	n. 密码，口令	3A
payload	[ˈpeiˌləud]	n. 有效载荷	9A
perceive	[pəˈsiːv]	v. 感知，感到	1A
performance	[pəˈfɔːməns]	n. 履行，执行，性能	4B

单词	音标	意义	课次
periodic	[piəri'ɔdik]	adj. 周期的，定期的	9B
peripheral	[pə'rifərəl]	adj. 外围的 n. 外围设备	2A
perish	['periʃ]	vi. 毁灭，死亡	1B
permanent	['pə:mənənt]	adj. 永久的，持久的	3B
permanently	['pə:mənəntli]	adv. 永存地，不变地	6A
personality	[,pə:sə'næliti]	n. 人格，人物，人名	1A
perspective	[pə'spektiv]	n. 观点，看法，观察	11B
pertinent	['pə:tinənt]	adj. 有关的，相干的，中肯的	10B
pervasive	[pə:'veisiv]	adj. 普遍深入的	1A
pharmaceutical	[,fɑ:mə'sju:tikəl]	n. 药物 adj. 制药(学)上的	1B
phenomenon	[fi'nɔminən]	n. 现象	8B
philosophical	[filə'sɔfikəl]	adj. 哲学的；达观的	10A
photosensor	[,fəutəu'sensə]	n. 光敏元件，光敏器件，光传感器	5A
piconet	['pikənet]	n. 微微网	7B
piezoelectric	[pai,i:zəui'lektrik]	adj. 压电的	4A
pill	[pil]	n. 药丸	1B
pioneer	[,paiə'niə]	n. 先驱，倡导者，先锋	5A
pitch	[pitʃ]	vt. 定位于，设定	5A
plaster	['plɑ:stə]	vt. 在……上大量粘贴；贴满	6B
platform	['plætfɔ:m]	n. 平台	10B
pollutant	[pə'lu:tənt]	n. 污染物质	8A
pollution	[pə'lu:ʃən]	n. 污染	11A
polymer	['pɔlimə]	n. 聚合体	3A
popular	['pɔpjulə]	adj. 流行的，受欢迎的	1A
popularity	[,pɔpju'læriti]	n. 普及，流行	5B
portability	[,pɔ:tə'biləti]	n. 可携带，轻便	
portal	['pɔ:təl]	n. 门户，入口	10B
position	[pə'ziʃən]	vt. 安置，决定……的位置	6A
positively	['pɔzətivli]	adv. 断然地，肯定地	11A
possess	[pə'zes]	vt. 拥有，持有，支配	11A
potentially	[pə'tenʃəli]	adv. 潜在地	10A
practically	['præktikəli]	adv. 实际上，事实上，在实践上 adv. 几乎，简直	5B
precaution	[pri'kɔ:ʃən]	n. 预防，警惕，防范	7A
precise	[pri'sais]	adj. 精确的，准确的	1A

单词	音标	意义	课次
precisely	[pri'saisli]	adv. 正好	6A
predefine	['pri:di'fain]	vt. 预先确定	1A
predict	[pri'dikt]	v. 预知，预言，预报	5A
prediction	[pri'dikʃən]	n. 预言，预报	1B
predominant	[pri'dɔminənt]	adj. 卓越的，支配的，主要的，突出的，有影响的	8A
preference	['prefərəns]	n. 偏爱，优先选择	4A
preliminary	[pri'liminəri]	adj. 预备的，初步的	11A
prerequisite	['pri:'rekwizit]	n. 先决条件　adj. 首要必备的	1A
presence	['prezns]	n. 存在，到场，出席	2B
presumably	[pri'zju:məbəli]	adv. 推测起来，大概	9A
prevent	[pri'vent]	v. 防止，预防	4B
primitive	['primitiv]	adj. 简单的，粗糙的	8A
printer	['printə]	n. 打印机	7A
privacy	['praivəsi]	n. 秘密	1A
proactive	[,prəu'æktiv]	adj. 积极主动的，主动出击的	1B
process	[prə'ses]	n. 过程，方法，程序，步骤 vt. 加工，处理	11B
producer	[prə'dju:sə]	n. 生产者，制作者	1B
productivity	[,prɔdʌk'tiviti]	n. 生产力	4B
profile	['prəufail]	v. 分析，介绍	5A
promising	['prɔmisiŋ]	adj. 有希望的，有前途的	1A
propagation	[,prɔpə'geiʃən]	n. (声波，电磁辐射等)传播	8A
property	['prɔpəti]	n. 性质，特性	8A
proponent	[prə'pəunənt]	n. 建议者，支持者	5B
propose	[prə'pəuz]	vt. 计划，建议，向……提议	10B
proprietary	[prə'praiətəri]	adj. 私有的，专有的　n. 所有者，所有权	8A
proprietor	[prə'praiətə]	n. 所有者，经营者	5A
protect	[prə'tekt]	vt. 保护，关税保护，投保	2B
protocol	['prəutəkɔl]	n. 协议	2A
provide	[prə'vaid]	v. 供应，供给	2A
pseudonym	['(p)sju:dənim]	n. 假名	3B
publication	[,pʌbli'keiʃən]	n. 出版物，出版，发行，发表	1A
publish	['pʌbliʃ]	v. 出版，刊印，公布，发表	8A
pulse	[pʌls]	n. 脉冲	4A

单词	音标	意义	课次
puppy	[ˈpʌpi]	n. 小狗，小动物	6B
purchase	[ˈpəːtʃəs]	vt.& n. 买，购买	1B,4B
quantify	[ˈkwɔntifai]	vt. 确定数量，量化	8A
quantity	[ˈkwɔntiti]	n. 量，数量	6B
quasi	[ˈkwɑːzi(ː), ˈkweisai]	adj. 类似的，准的	7B
query	[ˈkwiəri]	v. 询问，查询 n. 质问，询问，怀疑，疑问 v. 询问	1A,6A
queue	[kjuː]	n. 队列 vi. 排队	
radically	[ˈrædikəli]	adv. 根本上	9A
random	[ˈrændəm]	n. 随意，任意 adj. 任意的，随便的，随即的	3B
ratify	[ˈrætifai]	vt. 批准，认可	7A
ratio	[ˈreiʃiəu]	n. 比，比率	5B
raw	[rɔː]	adj. 未加工的，处于自然状态的	11B
react	[riˈækt]	vi. 起反应，起作用	1A
reaction	[ri(ː)ˈækʃən]	n. 反应	3A
realm	[relm]	n. 领域	3B,6B
recall	[riˈkɔːl]	vt. & n. 召回	1B
receiver	[riˈsiːvə]	n. 接收器，收信机	4A
recharge	[ˈriːˈtʃɑːdʒ]	vt. 再充电	11A
reconnect	[ˌriːkəˈnekt]	v. 再连接	7A
reconstructed	[ˌriːkənˈstrʌktid]	adj. 重建的，改造的	5B
recovery	[riˈkʌvəri]	n. 恢复	
rectangle	[ˈrektæŋgl]	n. 长方形，矩形	5A
rectangular	[rekˈtæŋgjulə]	adj. 矩形的，成直角的	6A
rectification	[ˌrektifiˈkeiʃən]	n. 整流	
redirect	[ˈriːdiˈrekt]	vt. 使改道，使改变方向，重寄	4A
redundancy	[riˈdʌndənsi]	n. 冗余	10A
redundant	[riˈdʌndənt]	adj. 多余的，冗余的	8A
refill	[ˈriːˈfil]	v. 再装满，补充，再充填 n. 新补充物，替换物	1B
reflect	[riˈflekt]	v. 反射，反映，表现	5A
refuse	[riˈfjuːz]	vt. 拒绝，谢绝 n. 废物，垃圾	5B
registration	[ˌredʒisˈtreiʃən]	n. 注册，报到，登记	5A

单 词	音 标	意 义	课次
registry	['redʒistri]	n. 注册，登记	3B
regrettably	[ri'gretəbli]	adv. 抱歉地，遗憾地，可悲地	1B
regulatory	['regjulətəri]	adj. 调整的	11B
reinvent	[,ri:in'vent]	vt. 彻底改造，重新使用	6B
relay	['ri:lei]	v. 转播 n. 继电器	9A
relevant	['relivənt]	adj. 有关的，相应的	1A
reliability	[ri,laiə'biliti]	n. 可靠性	
reliable	[ri'laiəbl]	adj. 可靠的，可信赖的	4B
remain	[ri'mein]	vi. 保持，逗留，剩余	2A
remote	[ri'məut]	adj. 远程的，遥远的	10A
remove	[ri'mu:v]	vt. 消除，删除，去除	10A
render	['rendə]	vt. 实施 vi. 给予补偿	1B
reopen	['ri:'əupən]	v. 重开，再开始	4B
repackage	[ri'pækidʒ]	vt. 重新包装	3B
replaceable	[ri'pleisəbl]	adj. 可代替的	3B
replacement	[ri'pleismənt]	n. 代替者，复位，交换，置换，移位	1B
repository	[ri'pɔzitəri]	n. 储藏室，仓库	9B
represent	[,ri:pri'zent]	vt. 表现，描绘	5A
representation	[,reprizen'teiʃən]	n. 表示法，表现	1A
reprice	['riprais]	vt. 重新定价	5A
reputation	[,repju(:)'teiʃən]	n. 名誉，名声	10A
request	[ri'kwest]	vt. & n. 请求，要求	2A
reset	['ri:set]	v. 复位	
resilient	[ri'ziliənt]	adj. 能复原的，有弹性的，有弹力的	3A
resistance	[ri'zistəns]	n. 反抗，抵抗，阻力，电阻，阻抗	3A
resource	[ri'sɔ:s]	n. 资源	4B
respectively	[ri'spektivli]	adv. 分别地，各个地	7B
response	[ris'pɔns]	n. 回答，响应，反应	2A
responsibility	[ris,pɔnsə'biliti]	n. 责任，职责	10A
restock	['ri:'stɔk]	vt. 重新进货，再储存	6B
restricted	[ris'triktid]	adj. 受限制的，有限的	9A
restriction	[ris'trikʃən]	n. 限制，约束	9A
restructure	[ri'strʌktʃə]	vt. 更改结构，重建构造，调整，改组	11A
retailer	[ri:'teilə]	n. 零售商人；传播的人	5A
retrievable	[ri'tri:vəbl]	adj. 可获取的	3B

单词	音标	意义	课次
retrieval	[ri'tri:vəl]	n. 取回，恢复，修补，重获	3A
retrieve	[ri'tri:v]	n. 检索；恢复，取回	2A
reusability	[ri,ju:zə'biləti]	n. 可重用性	
revision	[ri'viʒən]	n. 修订，修改，修正，修订本	9A
rice-sized	[rais-saizd]	adj. 米粒大小的	6A
roadmap	['rəudmæp]	n. 路标	11A
robust	[rə'bʌst]	adj. 健壮的，精力充沛的	9B
robustness	[rə'bʌstnis]	n. 鲁棒性，稳健性，健壮性	8A
rosy	['rəuzi]	adj. 蔷薇色的，玫瑰红色的	6B
rotating	[rəu'teitiŋ]	adj. 旋转的	8B
route	[ru:t]	n. 路线，路程，通道 v. 发送	2A
router	['rautə]	n. 路由器	2A
routine	[ru:'ti:n]	n. 日常事务，例行公事，常规，程序	10B
routing	['ru:tiŋ]	n. 行程安排，邮件路由	3A
sacrifice	['sækrifais]	n. & v. 牺牲	4B,9A
salespeople	['seilz,pi:pl]	n. 售货员，店员	4B
sample	['sæmpl]	n. 标本，样品，样值 vt. 取样，采样	
satellite	['sætəlait]	n. 人造卫星	2A
scalability	[,skeilə'biliti]	n. 可量测性	3B
scalable	['skeiləbl]	adj. 可升级的	2A
scan	[skæn]	v. 浏览，扫描 n. 扫描	5A
scanner	['skænə]	n. 扫描器	5A
scarce	[skɛəs]	adj. 缺乏的，不足的，稀有的，不充足的	8A
scatternet	['skætənet]	n. 分布网，分散网	7B
scenario	[si'nɑ:riəu]	n. 情景，情节	1A
scrambler	['skræmblə]	n. 干扰器，扰频器	
screening	['skri:niŋ]	n. 筛选，屏蔽	2B
scrub	[skrʌb]	v. 洗擦，擦净	5A
seamless	['si:mlis]	adj. 无缝的	10B
search	[sə:tʃ]	n. & v. 搜索，搜寻	3A
seasonal	['si:zənl]	adj. 季节性的，周期性的	5A
secure	[si'kjuə]	adj. 安全的，可靠的	2B
security	[si'kjuəriti]	n. 安全	1A
self-explanatory	[self-ik'splænətəri]	adj. 自明的，自解释的	7A
self-referenced	[self-'refərenst]	adj. 自引用的	1A

单词	音标	意义	课次
semantic	[si'mæntik]	adj. 语义的	11A
semantics	[si'mæntiks]	n. 语义学	8A
sense	[sens]	vt. 感知，感到，认识	1A
sensitive	['sensitiv]	adj. 敏感的，灵敏的	11B
sensitivity	['sensi'tiviti]	n. 灵敏度，灵敏性	7B
sensor	['sensə]	n. 传感器	1A
sequence	['si:kwəns]	n. 次序，顺序，序列	3B
server	['sə:və]	n. 服务器	10A
shape	[ʃeip]	n. 外形，形状 vi. 成形	6A
share	[ʃɛə]	n. 共享，参与，一份，部分，份额，参股 vt. 分享，均分，共有，分配 vi. 分享	4B
shelf	[ʃelf]	n. 架子；货架；书架	6B
shipment	['ʃipmənt]	n. 装船，出货	5A
shipper	['ʃipə]	n. 托运人，发货人	4B
shipping	['ʃipiŋ]	n. 海运，运送，航行	5B
shoplifter	['ʃɔpliftə]	n. 商店扒手	5A
shoplifting	['ʃɔpliftiŋ]	n. 入店行窃	5A
shortage	['ʃɔ:tidʒ]	n. 不足，缺乏	1B
signalling	['signəliŋ]	n. 信令，发信号	
significance	[sig'nifikəns]	n. 意义，重要性	5A
simplicity	[sim'plisiti]	n. 简单，简易	5A
simulation	[,simju'leiʃən]	n. 仿真，模拟	8A
simultaneous	[,siməl'teinjəs]	adj. 同时的，同时发生的	1A,6A
simultaneously	[siməl'teiniəsli]	adv. 同时地	7A
site	[sait]	n. 站点	4B
skim	[skim]	v. 快读，浏览，撇去	6B
slice	[slais]	n. 薄片，切片，一份，部分，片段	5A
smartphone	['smɑ:tfəun]	n. 智能电话	2A
smoothly	['smu:ðli]	adv. 平稳地	4B
smudge	[smʌdʒ]	n. 污迹 v. 弄脏，染污	5B
sniffer	[snifə]	n. 嗅探器	6A
solution	[sə'lju:ʃən]	n. 解答，解决办法，解决方案	5B
sophisticated	[sə'fistikeitid]	adj. 复杂的，先进的；老练的，富有经验的	11B

单词	音标	意义	课次
spacing	['speisiŋ]	n. 间隔,间距	5A
span	[spæn]	n. 跨度,跨距,范围 v. 横越	4B
sparse	[spɑːs]	adj. 稀疏的	3B
spatially	['speiʃəli]	adv. 空间地	8A
specific	[spi'sifik]	adj. 详细而精确的,明确的,特殊的	1A
specification	[,spesifi'keiʃən]	n. 详述,规格,说明书,规范	1A
spectral	['spektrəl]	adj. 光谱的,频谱的	10B
spectrum	['spektrəm]	n. 光谱,频谱,型谱	6B,9A
speedy	['spiːdi]	adj. 快的,迅速的	5B
split	[split]	v.(使)裂开,分裂,分离	7A
spoof	[spuːf]	v. 哄骗	7A
sporadic	[spə'rædik]	adj. 零星的,孤立的	3B
squelch	[skweltʃ]	v. 静噪	
stability	[stə'biliti]	n. 稳定性	1B
stack	[stæk]	n. 堆,一堆,堆栈 v. 堆叠	3A
staff	[stɑːf]	n. 全体职员	4B
stage	[steidʒ]	n. 发展的进程、阶段或时期	1A
standard	['stændəd]	n. 标准,规格 adj. 标准的,权威,第一流的	4B
starter	['stɑːtə]	n. 起动器	5B
state	[steit]	n. 情形,状态	1A
static	['stætik]	adj. 静态的,静力的	2A
station	['steiʃən]	n. 基站	10B
status	['steitəs]	n. 情形,状况	4A
steganography	[,stegən'ɔgrəfi]	n. 信息隐藏,速记式加密	5A
stimulate	['stimjuleit]	v. 刺激,激励	1B
stimulus	['stimjuləs]	n. 刺激物,促进因素,刺激	9B
stock	[stɔk]	n. 库存,原料	1A
stonewall	[stəunwɔːl]	vi. 妨碍,阻碍	5B
storage	['stɔridʒ]	n. 存储	10A
stream	[striːm]	n. 流,一串 v. 流,涌	11B
streamlined	['striːmlaind]	adj. 最新型的,改进的	10A
stressful	['stresful]	adj. 产生压力的,使紧迫的	10A
sublayer	['sʌb'leiə]	n. 子层,下层,次层	9A
submarine	['sʌbməriːn]	n. 潜水艇,潜艇	4B

单 词	音 标	意 义	课次
submicron	[ˈsʌbˈmaikrɔn]	adj. 亚微细粒的，亚微型的	3A
subnetwork	[sʌbˈnetwə:k]	n. 子网	2B
subsequent	[ˈsʌbsikwənt]	adj. 后来的，并发的	11A
subsequent	[ˈsʌbsikwənt]	adj. 后来的，并发的	2B
subsidiary	[səbˈsidjəri]	adj. 辅助的，补充的	1A
sub-signal	[sʌb-ˈsignl]	n. 子信号	7A
substitution	[ˌsʌbstiˈtju:ʃən]	n. 代替，取代作用，置换	5A
successive	[səkˈsesiv]	adj. 继承的，连续的	9A
sufficient	[səˈfiʃənt]	adj. 充分的，足够的	3B
suitability	[ˌsju:təˈbiləti]	n. 合适，适当，相配，适宜性	9A
suitable	[ˈsju:təbl]	adj. 适当的，相配的	11B
superframe	[ˈsju:pəˌfreim]	n. 超帧	9A
supplant	[səˈplɑ:nt]	vt. 排挤掉，代替	5A
supplier	[səˈplaiə]	n. 供应者，供应商，供给者	4B
surveillance	[səˈveiləns]	n. 监视，监督	11A
surveillance	[səˈveiləns]	n. 监视，监督	8A
swap	[swɔp]	v. & n. 交换	5A
sweep	[swi:p]	v. 扫，扫过，掠过	5A
swipe	[swaip]	vt. 刷……卡	6A
switch	[switʃ]	n. 交换机	11B
symbology	[simˈbɔlədʒi]	n. 符号学	5A
synchronization	[ˌsiŋkrənaiˈzeiʃən]	n. 同步	3A
synchronous	[ˈsiŋkrənəs]	adj. 同步的	
syntactic	[sinˈtæktik]	adj. 句法，语法	1A
tabulate	[ˈtæbjuleit]	v. 把……制成表格，以表格形式排列，列表	6B
tag	[tæg]	n. 标签，标识	1A
tamper	[ˈtæmpə]	v. 篡改	3A
tap	[tæp]	v. 轻打，轻敲	6B
target	[ˈtɑ:git]	n. 目标	9B
task	[tɑ:sk]	n. 任务，作业 v. 分派任务	2A
taste	[teist]	v. 品尝，辨味 n. 味道，味觉	1B
tattoo	[təˈtu:]	n. 纹身	6A
taxicab	[ˈtæksikæb]	n. 出租车	5B
tech-savvy	[tek-ˈsævi]	adj. 有技术头脑的，科技通	5B
telecom	[ˈtelekɔm]	n. 电信	7B

单词	音标	意义	课次
telecontrol	['telikən'trəul]	n. 遥控	
teleport	['telipɔ:t]	vt. 传送	4A
terabit	['terəbit]	n. 兆兆位(量度信息单位)	2A
term	[tə:m]	n. 术语	2B
terminal	['tə:minl]	n. 终端	3A
territorialisation	['teritɔ:riəlai'zeiʃne]	n. 按地区分配	11A
thermostat	['θə:məstæt]	n. 自动调温器,温度调节装置	7B
threaten	['θretn]	vt. 恐吓,威胁	11B
threshold	['θreʃəuld]	n. 门槛;阈值;起点,开端	2B
throughput	['θru:put]	n. 生产量,生产能力,吞吐量	11B
timeline	['taimlain]	n. 时间轴,时间线;大事年表	4A
tollbooth	['təulbu:θ]	n. 过路收费亭	6B
topology	[tə'pɔlədʒi]	n. 拓扑,布局	8A
toss	[tɔs]	v. 投,掷	6B
traceability	[,treisə'biləti]	n. 可追溯,可描绘,可描写	1B
traceable	['treisəbl]	adj. 可追踪的,起源于	1B
track	[træk]	n. 轨迹,跟踪,航迹,途径 vt. 追踪	4A
tradeoff	[treidɔf]	n. 折衷,权衡,(公平)交易	9A
traffic	['træfik]	n. 流量,通信量	2A
tragedy	['trædʒidi]	n. 悲剧,惨案,悲惨,灾难	3A
trail	[treil]	vt. 跟踪,追踪,拉,拖 n. 踪迹,痕迹,形迹	9A
transceiver	[træn'si:və]	n. 收发器	7B
transfer	[træns'fə:]	n.& vt. 移动,传递,转移,转账,转让	4A
transmission	[trænz'miʃən]	n. 播送,发射,传输,转播	1B,2A
transmit	[trænz'mit]	vt. 传输,发射,传播 vi. 发射信号,发报	2A
transmitter	[trænz'mitə]	n. 传导物,发报机,发射机	4A
transponder	[træn'spɔndə]	n. 变换器,转调器;异频雷达收发机	6A
transport	[træns'pɔ:t]	vt. 传送,运输	3B
treatment	['tri:tmənt]	n. 待遇,对待,处理	3B
tremendous	[tri'mendəs]	adj. 极大的,巨大的	11B
trigger	['trigə]	vt. 引发,引起,触发	8B
trilateration	[trai,lætə'reiʃne]	n. 三边测量(术)	4A

单 词	音 标	意 义	课次
truck	[trʌk]	n. 卡车，交易，交换 vt. 交易，交往，以卡车运输 vi. 驾驶卡车，以物易物	4B
tuning	[ˈtjuːnɪŋ]	n. 调谐	
tunnel	[ˈtʌnl]	n. 通道，隧道，地道	4B
ubiquitous	[juːˈbɪkwɪtəs]	adj. 泛在的，到处存在的，普遍存在的	4A
ultrasonic	[ˌʌltrəˈsɔnɪk]	adj. 超音速的，超声的 n. 超声波	4A
unattended	[ˌʌnəˈtendɪd]	adj. 没人照顾的，未被注意的	8A
unattractive	[ˌʌnəˈtræktɪv]	adj. 不吸引人的	10B
unchain	[ʌnˈtʃeɪn]	vt. 解除，释放	4A
unclear	[ʌnˈklɪə]	adj. 不清楚的	11A
underestimate	[ˈʌndərˈestɪmeɪt]	vt.& n. 低估	1B
undergo	[ˌʌndəˈɡəʊ]	vt. 经历，遭受，忍受	1A
undoubtedly	[ʌnˈdaʊtɪdli]	adv. 毋庸置疑地，的确地	11B
unfortunately	[ʌnˈfɔːtjunətli]	adv. 不幸地	5B
unidirectional	[ˌjuːnɪdɪˈrekʃənəl]	adj. 单向的	
unified	[ˈjuːnɪfaɪd]	adj. 统一的，统一标准的，一元化的	1B
unique	[juːˈniːk]	adj. 唯一的，独特的	11A
uniquely	[juːˈniːkli]	adv. 独特地，唯一地	1A
universal	[ˌjuːnɪˈvɜːsəl]	adj. 普遍的，全体的，通用的，宇宙的，世界的	5A
universality	[ˌjuːnɪvəˈsælɪti]	n. 普遍性，一般性，多方面性，广泛性	5A
universe	[ˈjuːnɪvɜːs]	n. 宇宙，世界，万物，领域	11B
unlicensed	[ˈʌnˈlaɪsənst]	adj. 没有执照的，未经当局许可的，无节制的	9A
unlimited	[ʌnˈlɪmɪtɪd]	adj. 无限的，无约束的	3B
unobtrusive	[ˌʌnəbˈtruːsɪv]	adj. 不引人注目的	7A
unobtrusiveness	[ˌʌnəbˈtruːsɪvnɪs]	adj. 不突出的，不引人注意的	3B
unoccupied	[ʌnˈɔkjupaɪd]	adj. 没有人住的，无人占领的，空闲的	4A
unreachable	[ʌnˈriːtʃəbl]	adj. 取不到的，不能得到的	10A
unrest	[ʌnˈrest]	n. 不安定，动荡，骚乱	1B
update	[ʌpˈdeɪt]	v. & n. 更新	3A
upload	[ˈʌpˌləud]	v. 上传，上载	10B
uprising	[ʌpˈraɪzɪŋ]	n. 暴动；升起	1B
utilise	[ˈjuːtɪlaɪz]	vt. 利用	11A
vague	[veɪɡ]	adj. 含糊的，不清楚的	7B

单 词	音 标	意 义	课次
valid	['vælid]	adj. 有效的，有根据的，正确的	9A
validation	[,væli'deiʃən]	n. 确认	9A
valuable	['væljuəbl]	adj. 有价值的	11B
variable	['vɛəriəbl]	n. 变数，变量 adj. 可变的，易变的，变量的	1A
vector	['vektə]	n. 向量，矢量 vt. 无线电导引	9B
vegetation	[,vedʒi'teiʃən]	n. 植物，草木	8B
vendor	['vendɔː]	n. 卖主	6A
vent	[vent]	n. 通风口	8B
vibration	[vai'breiʃən]	n. 振动，颤动，摇动，摆动	8A
vice versa	['vaisi 'vəːsə]	adv. 反之亦然	5A
videoconference	[,vidiəu'kɔnfərəns]	n. 视频会议	4A
vintage	['vintidʒ]	adj. 古老的，最佳的，过时的	6A
virtual	['vəːtjuəl]	adj. 虚拟的，实质的	1A
virtually	['vəːtjuəli]	adv. 事实上，实质上；几乎	1A
volcano	['vɔukeinəu]	n. 火山	8B
voluntary	['vɔləntəri]	adj. 自动的，自愿的，主动的，故意的	5A
vulnerability	[,vʌlnərə'biləti]	n. 弱点，攻击	7A
waggle	['wægl]	v. 来回摇动，摆动	9B
warehouse	['wɛəhaus]	n. 仓库，货栈，大商店 vt. 存入仓库	4B
wastewater	['weistwɔːtə]	n. 废水	8B
wavelength	['weivleŋθ]	n. 波长	7B
well-defined	['weldifaind]	adj. 定义明确的，明确的	3A
wholesaler	['həulseilə]	n. 批发商	1B
widespread	['waidspred]	adj. 分布广泛的，普遍的	7A
wireless	['waiəlis]	adj. 无线的	10B
withstand	[wið'stænd]	vt. 抵挡，经受住	8A
workload	['wəːkləud]	n. 工作量	10A
workstation	['wəːksteiʃən]	n. 工作站	4A
worldwide	['wəːldwaid]	adj. 全世界的	4B
worry	['wʌri]	n. 烦恼，忧虑 vt. 使烦恼，使焦虑；困扰，折磨	10A
zone	[zəun]	n. 地域，环带，圈 vt. 环绕，使分成地带 vi. 分成区	4A

词 组 表

词 组	意 义	课次
"thing" to "thing"	"物"到"物"	3A
a grain of dust	一粒灰尘	8A
a leg of lamb	羊腿	6A
a pack of	一包，一盒	5B
a pair of	一对	9A
a pop	每件值……钱	5B
a series of	一连串的	2A
a suite of	一套，一系列	9B
a variety of	多种的	5B
absolute gain	绝对增益	
Absolute Phase Shift Keying (APSK)	绝对相移键控	
Absolute RF Channel Number (ARFCN)	绝对射频信道号	
access channel	入网信道	
access mode	存取方式，访问方式	9B
actionable information	全面、精确的信息	3A
active mode	活动模式	
active RFID tag	有源 RFID 标签	6A
active substance	有效物质，有效成分	1B
active tag	有源标签，主动标签	3A
Ad hoc Network	自组织网络	9A
address space	地址空间	1A
adjacent-channel power	邻道功率	
advice of charge	计费信息	
Agent-based Modeling and Simulation	基于代理的建模和仿真	8A
agricultural industry	农业	8B
ambient intelligence	环境智能	1A
ambient noise	环境噪声	
amplitude detection	幅度检波	
amplitude distortion	幅度失真	
amplitude limiting circuit	限幅电路	
Amplitude Modulation (AM)	幅度调制	
amplitude shift keying	幅移键控，幅变调制，振幅偏移键控法	9A
analog telephone system	模拟电话系统	4B
analogue circuit	模拟电路	3A

词 组	意 义	课次
antenna directivity diagram	天线方向性图	
antenna gain	天线增益	
area monitoring	区域监测	8B
arrive at	到达	2B
Artifical Intelligence (缩写为 AI)	人工智能	1A
as long as	只要，在……的时候	4A
at full capacity	满功率，满负载	10A
at lower costs	以较低价格，以较低成本	4B
at the threshold of	在……的开始	2B
attach to	把……放在，附加到……	1A
authentication certificate	证书	2B
automatic channel selection	自动信道选择	
Automatic Dialing Unit (ADU)	自动拨号设备	
Automatic Frequency Control (AFC)	自动频率控制	
Automatic Frequency Fine Control (AFEC)	自动频率微调	
Automatic Gain Control (AGC)	自动增益控制	
Automatic Power Control (APC)	自动功率控制	
Automatic Repeat Request (ARQ)	自动请求重发	
automatic switching equipment	自动交换设备	
autonomous control	自主控制	1A
available power	资用功率	
average envelope detection	平均包络检波	
average power	平均功率	
back end	后端	10A
back up	备份	4A
background noise	背景噪声	
balanced code	平衡码	
band allocation	频带划分	
bar code	条形码	1B
barcode reader	条码阅读机	5A
bargain hunter	买便宜货的人，投机商人	5B
base on	基于	1A
base station	基站，基地	8B
Basic Rate (BR)	基本速率	7B
battery capacity	电池容量	

词 组	意 义	课次
battery pack	电池组	8B
be about to	将要，正打算	5B
be adhered to	应坚持	5A
be affixed to	被贴到	6A
be associated with	和……联系在一起；与……有关	6A
be aware of	知道	6B
be composed of…	由……组成	8A
be confined to	限制在，局限于	10A
be defined as	被定义为	1A
be enabled to	使能够	1A
be equipped with	装备	10B
be going to	要，会，将要	10B
be integrated into	统一到……中，整合到……中	1A
be intended to be	规定为，确定为	9B
be involved in	涉及，专心	1A
be kept in mind	牢记在心	11B
be located at	位于	2B
be meant to	有意要，打算	9A
be of no use	无用	8A
be prone to	有……的倾向，易于	8A
be referred to	被提及；涉及	5A
be replaced by	被代替	6B
be responsible for	为……负责，形成……的原因	4B
become effective	生效	7B
best route	最佳路由	2B
bi-directional link	双向链接	7B
big data	大数据	11B
bill of materials	材料单	3B
binary code	二进制码	
Binary Different Phase Shift Keying (BDPSK)	二相差分相移键控	
Binary Frequency Shift Keying (BFSK)	二进制频移键控	
binary keying	二进制键控	9A
Binary Phase Shift Keying (BPSK)	二相相移键控	
bit error rate	比特误码率	

词 组	意 义	课次
bit rate	比特率	
bit stream	比特流	
block code	分组码	
boarding pass	登机证	5A
bottom-up made	自底向上的	1A
branch office	分支机构	4B
break down	毁掉，停顿，中止，垮掉	10A
build up	增进，增大	5A
build upon	指望，依赖，建立于	9B
bulk product	散货	3B
bus network	总线网络	2B
business cluster	产业聚集	11A
by default	默认	7A
by means of	依靠，用，借助于	9A
call for	要求，提倡；为……叫喊，为……叫	5B
capable of	有……能力的，可……的	5B
capacitively coupled tag	电容耦合标签	6B
capital expenditure	资本支出，基本建设费用	10B
capital goods	生产资料；资本财货，资本货物	1B
car recharging station	汽车充电站	11A
carrier frequency offset	载波频率偏置	
carrier power	载波功率	
Carrier Sense Multiple Access (CSMA)	载波检测多址	
catch on with	迎合，跟随	6B
CDMA cellular system	码分多址蜂窝系统	
CDMA PCN	码分多址个人通信网	
cellular system	蜂窝系统	
centralised architecture	集中式体系结构	8A
certainty factor	可信度	
channel assignment	信道指配	
channel bandwidth	频道带宽，通道带宽	10B
channel capacity	信道容量	
channel coding	信道编码	
channel efficiency	信道效率	
channel gate	信道门	

词 组	意 义	课次
channel selective mode	信道选择方式	
channel spacing	信道间隔	
channel time slot	信道时隙	
check out	结账；检验，合格	1B
checkout station	结账台	6A
chewing gum	口香糖	5A
Chirp Spread Spectrum (CSS)	线性调频技术	9A
chop up	切开，切细	7B
client software	客户软件	4B
climate change	气候变化	11A
clock tick	时钟节拍，时钟周期	7B
closed loop	闭环	1A
closed loop control	闭环控制	
cloud computing	云计算	10A
cloud computing technology	云计算技术	3A
cloud-based service	基于云的服务	11A
cluster tree	簇状结构，簇状布局	9A
co-channel signalling	共道信令	
co-channel suppression	共道抑制	
code conversion	码交换	
code division	码分	
code division multiple access	码分多址	
code generator	编码发生器	5B
code word	码字	
coded signal	编码信号	
cognitive competence	认知能力，感知能力	11A
cognitive radio	认知无线电	3B
collective intelligence	群体智能	11A
collision avoidance	防止空中相撞	9A
come by	获取，得到；从旁经过，通过	5B
come up with	提出，想出，赶上	4A
commercial greenhouse	商业性暖房，商业性大棚	8B
commodity market	商品市场	1B
common man	平民	10B

词 组	意 义	课次
communication device	通信设备	8A
communication protocol	通信协议	3A
computer industry	计算机行业	10A
computer security	计算机安全	2B
Condition Based Maintenance (CBM)	状态维护，状态维修	8B
conductive carbon ink	导电碳墨	6B
conform to	符合，遵照	9B
congestion control	拥挤控制	
connect with	连接，联络	2A
consist of …	由……组成	3A
Continuous Phase-Frequency Shift Keying (CP-FSK)	连续相位频移键控	
converter conversion loss	变频损耗	
conveyer belt	输送带	1B
conveyor line	输送线	5A
convolutional code	卷积码	
cope with	成功应付，妥善处理	8A
core switch	中心交换	2B
coupled circuit	耦合电路	
coupled factor	耦合度	
coupling factor	耦合系数	
coupling loss	耦合损耗	
credit card	信用卡	6A
cross platform compatibility	跨平台兼容性	3B
cultural icon	文化符号，文化图腾，文化标志	10B
custom code	自定义码	5A
cycle-skipping	跳周	
cyclic code	循环码	
data acquisition circuit	数据采集电路	
data analysis program	数据分析程序	10A
data analytics	数据分析	11A
data cable	数据传输电缆	7B
data capture	数据捕捉	1A
data circuit	数据电路	
data collection	数据收集，数据汇集，收集资料	5A
data communication network	数据通信网络	

词 组	意 义	课次
data link layer	数据链路层	
data logging	数据资料记录	1A
data processing	数据处理	10A
data protection	数据保护	
data resources	数据资源	2B
data signal	数据信号	
data source	数据源	
data switching equipment	数据交换设备	
Data Terminal Equipment (DTE)	数据终端设备	
data theft	数据失窃	10B
data transfer	数据传送	
data transfer rate	数据传输率	10B
data transmission line	数据传输线路	
data-link level	数据链路层	2B
deal with	安排,处理,涉及	8A
debit card	借记卡	6B
decentralized processing	分散处理,分布处理	3A
decision maker	决策者	1A
dedicated channel	专用信道	
dedicated line	专用线	
deduct from	扣除	6B
default setting	默认设置	7A
deliver to	转交,交付,传达	4B
Department of Defense	(美国)国防部	5B
department store	百货公司	5A
desktop switch	桌面交换	2B
detection distortion	检波失真	
detection efficiency	检波效率	
development trend	发展趋势	3A
device type	设备类型	9B
dial in	拨号,拨入	2B
digilogue channel	数模信道	
digilogue circuit	数模电路	
digital camera	数码相机	4A
digital city	数字城市	11A

词 组	意 义	课次
digital domain	数字域	3B
Digital Earth	数字地球	1A
digital identity	数字身份	3B
digital integrated circuit	数字集成电路	
digital line path	数字有线通道	
digital path	数字通道	
digital phase-locked loop	数字锁相环	
digital processing	数字处理，数字加工	3A
digital radio path	数字无线通道	
digital radio section	数字无线段	
digital signal	数字信号	
Digital Signal Process (DSP)	数字信号处理	
digital storage device	数字化存储设备	10A
Digital to Analog (D/A) conversion	数/模变换	
direct distribution	直接分配	
Direct Sequence Spread Spectrum (DSSS)	直接序列扩频，直接序列扩频通信	9A
direct sequence Ultra-WideBand (UWB)	直接序列超宽带	9A
direct-coupled amplifier	直接耦合放大器	
distributed computing	分布计算	1A
distributed control architecture	分布控制式体系结构	8A
distributed parameter network	分布参数网络	
distribution link	分布路线	
door frame	门框	6A
dozens of	许多的	4B
drug allergy	药物过敏反应	5A
due to	由于	1A
duplex, full duplex	双工、全双工	
duty cycle	工作比	9B
economic restructuring	经济改革	11A
edge router	边式路由器	2B
Eethernet network	以太网	7B
electric shielding	电屏蔽	
electrical current	电流	6B
electrical meter	电子仪表	9B
Electro Magnetic Compatibility (EMC)	电磁兼容性	

词　组	意　义	课次
electromagnetic coupling	电磁耦合	
electromagnetic emission	电磁辐射	
electromagnetic environment	电磁环境	
electromagnetic field	电磁场	
electromagnetic induction	电磁感应	
electronic circuit	电子电路	8A
electronic community	电子社区	11A
element synchronism	码元同步	
embedded intelligence	嵌入智能	11A
empowerment intelligence	赋权智能	11A
end of transmission character	传输结束字符	
end point	端点	2A
end to end	端对端	3A
end to end communication	端对端通信	3B
end-of-message code	电文结束代码	
enemy intrusion	敌人入侵	8B
energy conservation	能源节约	10B
energy harvesting	能量收集，能量采集	8A
energy saving lamp	节能灯	11A
energy source	能源	8A
envelop delay distortion	包络延时失真	
environmental sensing	环境感应	8B
environmental sensor network	环境传感器网络	8B
equalization network	均衡网络	
error correcting capability	纠错能力	
error correcting code	纠错码	
error correction	纠错，误差修正	5B
error detecting code	检错码	
error detection	检错	
error rate	差错率	5A
even parity check	偶校验	
event handler	事件句柄；事件处理程序	8A
event transfer	事件传输	1A
event-driven architecture	事件驱动的体系结构	1A
event-driven programming model	事件驱动编程模型	8A

词 组	意 义	课次
exclusive-NOR gate	"异或非"门	
exclusive-OR gate	"异或"门	
expiration date	产品有效期	6B
fading bandwidth	衰落带宽	
fading envelope	衰落包络	
fading rate	衰落率	
failure rate	失效率	
family tree	系谱，系谱图，族谱，族谱图	3B
far-field region	远场区	
fashion company	时装公司	5B
Fast Moving Consumer Product	快速消费品	1A
fault tolerance	容错	8A
fault-tolerant technique	容错技术	
feedback control system	反馈控制系统	
filing clerk	文件管理员	4A
fill up	填补，装满	6B
fire brigade	消防队	8B
flash memory	闪存	9B
focus on	集中	1A
fool sb. into doing sth.	哄骗某人做某事	10A
for that matter	就此而言，而且，说到那一点	5B
Forward Error Control (FEC)	前向差错控制	
Forward Error Correction (FEC)	前向纠错	
forward signal	前向信号	
frequency band	频带	
frequency channel	频道	
Frequency Division Multiple Access (FDMA)	频分多址	
frequency hop	跳频	7A
Frequency Modulation (FM)	频率调制	
frequency multiplication	倍频	
frequency reuse	频率再用	
frequency shift	频移	
Frequency Hopping Spread Spectrum (FHSS)	跳频扩频	7B
front end	前段	10A
full open loop	全开环	1A

词　　组	意　　义	课次
Full Function Device (FFD)	全功能设备，全功能装置	9A
game theory	博弈论，对策论	3A
garage sale	现场旧货出售	5B
gateway server	网关服务器	2B
gather from…	从……获悉，从……收集	8A
Gaussian Frequency Shift Keying (GFSK)	高斯频移键控	7B
geometric pattern	几何图形	5A
gigabit per second	每秒吉字节	10B
give service to…	为……服务	9A
Global Positioning System (GPS)	全球定位系统	7B
global roaming	国际漫游	10B
go to waste	浪费掉，白费	10A
graphical user interface	图形用户界面	2B
gravity feed water system	重力给水系统	8B
grid computing system	网格计算系统	10A
hand-held computer	手持式计算机	7B
hard drive	硬盘驱动器	10A
hav a close relationship with	有密切的联系	4B
have access to	有权使用	2B
have no control over	不能控制	4B
health care	卫生保健	11A
hierarchical structure	层次结构，分级结构	3B
high frequency gain control	高频增益控制	
high layer protocol	高层协议	
high reliability	高可靠性	9B
high-frequency device	高频设备	6A
highpass filter	高通滤波器	
highway toll passcard	高速公路收费卡	6B
hold true	适用，有效	2A
hook point	钩入点	9A
human intelligence	人工智能	11A
human-made object	人造物体	1A
hypertext transfer protocol	超文本传输协议	2A
identity management	身份管理	11A
in a manner	在某种意义上；在一定程度上；以……方式	11A

词　　组	意　　义	课次
in a meaningful way	以有意义的方式	2A
in a nutshell	简单地，简约地	2A
in a way	在某种程度上，稍稍	3B
in all directions	四面八方，各方面	7A
in an attempt to	力图，试图	6B
in combination with …	与……结合	11B
in contrast	相反，大不相同	6A
in contrast with	和……形成对比，和……形成对照	9A
in nature	实际上，本质上	11B
in order to	为了……	3A
in pairs	成双地，成对地	5A
in relation to …	关于，涉及，与……相比	4A
in response to	响应，适应	11B
in reverse	反过来，与……相反	7A
in some cases	在某些情况下	2B
in somebody's name	以某人的名义	1B
in terms of …	根据……，按照……，用……的话，在……方面	3A
in the case of …	在……的情况	6A
in the construction of	建筑，建造	3B
in the field	在野外	4B
in the form of …	以……的形式	8A
in the middle of …	在……的中间	7A
in the vicinity	在附近	5A
in the vicinity of	在邻近	8A
in transit	运送中的	1B
in-band harmonic	带内谐波	
in-band noise	带内躁声	
in-band signalling	带内信令	
inductively coupled RFID tag	电感耦合 RFID 标签	6B
information fusion	信息融合	8A
information hopper	信息料斗，信息仓，信息存储池	4A
information quantity	信息量	
information sink	信宿	
information source	信源	

词 组	意 义	课次
information systems	信息系统	11A
Information Technological Equipment (ITE)	信息技术设备	
information technology	信息技术	
information theory	信息论	
information transfer	信息传递	
input port	输入端口	2B
instant checkout picture	即时检测图像	6B
instrumentation intelligence	仪器仪表智能化	11A
integrate with …	使与……结合，与……整合	2A
integrated circuit	集成电路	
integrated digital network	综合数字网	
Integrated Services Network (ISN)	综合业务网	
intelligent city	智能城市	11A
intelligent interface	智能接口	1A
interact with …	与……相互作用，与……相互影响；与……相互配合	3B
interface software	接口软件	10A
interference parameter	干扰参数	
interfering resource	干扰源	
interfering signal	干扰信号	
interfering suppression	干扰抑制	
internal antenna	内置天线	8A
internal impedance	内阻	
International Society of Automation	美国国际自动化协会	8A
Internet Engineering Task Force(IETF)	互联网工程任务组	8A
Internet of Things(IOT)	物联网	1A
Internet Protocol (IP)	因特网协议	1A
interworking between network	网间互通	
invest in	投资于，买进	6A
irrigation automation	自动灌溉	8B
jigsaw puzzle	七巧板，智力拼图玩具	2A
jump in	投入	5B
keep track of	记录，持续追踪；与……保持联系	4A,3B
keep up with	跟上	6B
kind of	有点儿，有几分	2A

词 组	意 义	课次
laser scanner	激光扫描仪	5A
law enforcement official	执法人员	6A
law firm	律师事务所	10A
layer-3 switch	第三层交换	2B
lead code	前导码	
leased line	租用线,专用线	4B
life span	寿命,使用期限	6A
light beam	光束	5A
linear block code	线性分组码	
linear dimension	线性维度	1A
linear distortion	线性失真	
link protocol	链路协议	
lithium thionyl chloride battery	锂亚硫酰氯原电池	4A
Local Area Network (LAN)	局域网	2B
logic channel	逻辑信道	
logical circuit	逻辑电路	
logon procedure	登录规程	2B
long before	很早以前	8B
look at … as	把……看做	2A
look up	查表	2B
low pass filter	低通滤波器	
low-latency device	低延迟的装置	9A
machine learning	机器学习	3A
mad cow disease	疯牛病	1B
magnetic field	磁场	6B
magnetic strip	磁条,磁片	6A
mains supply	干线供电,市电电源,交流电源	9B
maintainability	可维修性	
make inroads in	成功进入,进入,入侵	5A
make sure	确保,确定;确信,证实	10A
make up	连成,弥补,缝制,整理	4B
make waves	兴风作浪,制造纠纷	5B
making inference	推论	8A
man-made interference	人为干扰	
man-made noise	人为噪声	

词 组	意 义	课次
margin for error	误差容许量，误差限度	5B
marked with	以……为标记，以……表明	1B
market analyst	市场分析者	1A
market mechanism	市场机制，市场调节职能，市场法则	3A
master clock	主时钟	7B
master-slave structure	主-从结构	7B
matching network	匹配网络	
Maximum Usable Frequency (MUF)	最高可用频率	
Mean Time Between Failures (MTBF)	平均故障间隔时间	
Mean Time To Failures (MTTF)	平均无故障时间	
Media Access Control(MAC)	介质访问控制	9A
median of signal level	信号电平中值	
medical history	病史	5A
meet demand	满足需求	2A
mesh network	网状网络	8A
message format	消息格式	
metropolitan city-region	都市圈	11A
minimum distance	最小码距	
Minimum Shift Keying (MSK)	最小频移键控	
misting system	喷雾系统	8B
mobile Internet	移动因特网	10B
mobile phone network	移动电话网络	8A
mobile robot	移动式机器人，移动式遥控装置	8A
Mobile Termination (MT)	移动终端	
mobile user	移动用户	2B
modulation characteristic	调制特性	
move forward	前进	1A
moving part	运动机件，可动部分	5A
MultiChannel Access (MCA)	多信道选取	
multi-layered security system	多层安全系统	11B
multiple timeframe	多重时帧	
multiplexor channel	多路信道	
multipoint access	多点接入	
name space	名空间	3B
narrowband disturbance	窄带干扰	

词 组	意 义	课次
narrowband emission	窄带发射	
network address	网络地址	3B
network bridge	网桥	2B
Network Control Center (NCC)	网络控制中心	
network interworking	网络互联	
network optimization	网络优化	
network path	网络通路	
network point	网点	2B
network resource	网络资源	
network segmentation	网段	11B
network switch	网络转接	2B
network topology	网络拓扑	
network transfer delay	网络传输时延	
neural network	神经元网络	
Near Field Communication(NFC)	近场通信，近距离通信	
nominal data rate	标称数据率	10B
nominal value	标称值	
nonliear distortion	非线性失真	
Non Return to Zero (NRZ)	非归零码	
nonsynchronized network	异步网	
nuclear waste	原子能工业废料	5A
Object Naming Service	物件名称解析服务	6B
occupied bandwidth	占用带宽	
offset quadrature phase shift keying	偏移四相移相键控	9A
oil pipeline	输油管	8B
on a large scale	大规模地	6B
on behalf of …	代表……	1B
on the cusp of	正在着手，正处于	6B
on the lookout for	寻找，注意	7A
on the shelf	在货架上，在搁板上，束之高阁，不再流行；滞销的	5B
on the side	另外	10A
on their own initiative	他们自己主动	1A
one another	互相，彼此	2A
one at a time	每次一个	6B

词　组	意　义	课次
Open Geospatial Consortium (OGC)	开放地理空间联盟	8A
open innovation	开放式创新	11A
operating system	操作系统	3A
orchestration intelligence	业务流程智能化	11A
output port	输出端口	2B
outside resources	外部资源	2B
over and over	反复，再三	6A
packet switching	分组交换	
packet-based protocol	基于分组的协议	7B
packet-switched network	包交换网络	2B
paint in	补画上去，加绘上去	6B
paradigmatic shift	范式变革	10B
parallel computing	并行计算	1A
Parallel，not Sequential Spread Spectrum (PSSS)	平行，而不是序列扩展频谱	9A
parity bit	奇偶校验位	
parity check code	奇偶校验码	
participate in	参加，参与，分享	3B
pass on	传递	9B
pass through	经过，通过	2A
passenger protocol	乘客协议	4B
passive RFID tag	无源 RFID 标签	6A
pattern recognition	模式识别	3A
pay a toll	付通行费	6B
peak power	峰值功率	
peak value	峰值	
peanut butter	花生酱	6B
peer network	对等网	3A
pending message	待审消息	9A
periodic check	定期检查	9A
Personal Area Network(PAN)	个人局域网	9A
Personal Communication Networks (PCN)	个人通信网	
Personal Identification Number (PIN)	个人识别号码	
phase control	相位控制	
phase deviation	相位偏移	
Phase Modulation (PM)	调相	

词组	意义	课次
Phase Shift Keying (PSK)	相移键控	
physical address	物理地址	2B
physical assets	实物资产	11A
physical infrastructure	物质基础设施	11A
physical layer	物理层	9A
physical site visit	实地考察	8B
pile on …	使堆积在……	5B
pilot project	(小规模)试验计划	1B
plaster on	在……上面涂抹或粘贴	6B
plug into	把(电器)插头插入，接通	4A
Point of Common Coupling (PCC)	公共耦合点	
point-of-sale management	销售点管理	5A
pollution control board	污染控制局	8B
portable equipment	可携带设备	7B
power consumption	能量消耗，功率消耗	7B
power converter	电力变换器，整流器	3A
power management	动力管理	9A
power plant	发电厂，发电站	11A
powerline networking	电力线组网	9B
Preassigned Multiple Access (PMA)	预分配多址	
predictive modelling	预测模型	11A
prevent … from	阻止；制止	2B
print out	打印出，印出，显示	4A
private data network	专用数据网	
probabilistic decoding	概率译码	
processing unit	处理器，处理部件	8A
program command	程序指令	
program optimization	程序优化	
programming language	程序设计语言	3A
propagation condition	传播条件	7B
protect from [against]	防止……遭受……；使……免于，保护……使其不受	2B
provide … with …	为……提供……	5B
pseudorandom code	伪随机码	
pseudorandom sequence	伪随机序列	

词　组	意　义	课次
public authority	政府当局	1B
public data network	公用数据网	
public key system	公开密匙体制	
public resource	公共资源	4B
pull off	努力实现，赢得	6B
pulse code	脉冲编码，脉码	
put … to use	开始使用	5B
put a new spin on	对……作出新的解释	4A
put emphasis on	强调	10B
put up	举起，抬起，进行，提供，表现出	5B
Quadrature Amplitude Modulation (QAM)	正交调幅	
quadrature frequency hopping	正交调频	
quality management	质量管理	1B
quest for	追求，探索	1B
queuing delay	排队延时	
queuing discipline	排队规则	
queuing model	排队模型	
Radio Frequency Interface (RFI)	射频干扰	
radio tag	无线标签，无线标识	1A
radio transceiver	无线电收发器	8A
radio transmitter	无线电广播发射机	4A
radio waves	无线电波	7A
random exponential backoff	随机指数回退	9A
reactive near-field region	感应近场区	
real life	现实生活，实际生活	11A
Real-Time Location Systems (RTLS)	实时定位系统	7B
real-time monitoring system	实时监控系统	11A
real-time stream processing	实时流处理	11B
recycle bin	回收站	6B
reduce cost	降低成本	11A
Reduced-Function Device (RFD)	功能简化设备	9A
refer to	涉及，关系到；提到，谈到	1A
regardless of	不管，不顾	5B
registered mail	挂号信，挂号邮件	5A
relative to	相对于	6A

词 组	意 义	课次
reliability and maintainability assurance	可靠性和维修性保证	
rely on	依赖，依靠	1A
remote access	远程访问	2B
repair shop	维修车间	6A
replaced … with …	用……替换……	4B
reset procedure	重置规程	
respond to …	对……做出响应	6A
REverse Control Channel (RECC)	反向控制信道	
ring network	环形网络	2B
round-robin fashion	轮流的方式	7B
router program	路由程序	2B
routing table	路由表	8A
run out of	用完	2A
sample rate	采样率	
sampling frequency	采样频率	
sampling period	采样周期	
satellite phone	卫星电话	8A
scan channel	信道扫描	
security policy	安全策略	2B
send out	发送	2B
separate … from	分离，分开	2B
serial communication	串行通信	7B
serial number	序列号，系列号	6B
serve the same purpose as	与……目的一样	6A
server virtualization	服务器虚拟化	10A
Service Set IDentifier (SSID)	服务集标识符	7A
session layer	会话层	
set of rules	规则组，规则集	2A
set up	设立，竖立，架起，升起	4A
Set-Top Box (STB)	机顶盒	7B
shell out	交付，支付	6B
shipping company	轮船公司	4B
shipping truck	运输卡车	4B
shop floor	车间，工场	1B
shopping cart	购物车	6B

词组	意义	课次
shopping mall	大型购物中心	6B
short distance	短距离	7B
short-range wireless technology	短距离无线技术	3B
shut down	放下，关下，(使)机器等关闭，停车	6B
sidechannel attack	侧信道攻击	3A
sign in	签到，签收	4B
Signal to Noise Ratio(SNR)	信噪比	
signalling interworking	信令互通	
signalling link	信令链路	
signalling message	信令消息	
signalling point	信令点	
silicon chip	硅片	6B
simplex operation	单工操作	
single hop	单一跳跃	3B
smart appliance	智能器具	9B
smart card	智能卡	11A
smart city	智慧城市	11A
smart dust	智能微尘	1B
smart meter	智能仪表	11A
smart metering	智能仪表	9B
smart object	智能物体，智能对象	1A
smart poster	智能海报	4A
smart space	智能空间	1A
social networking site	社交网站	10B
software defined radio	软件定义无线电	3B
solar panel	太阳能电池板	8B
solar power	太阳能	11A
sort out	挑选出	1B
Space Division Multiple Address (SDMA)	空分多址	
spatial scale	空间尺度，空间比例尺，空间等级	1A
spectral efficiency	频谱效率，频谱利用率	10B
spill the beans	泄露秘密，说漏嘴	6A
split … into …	把……分为……	7A
spread to	传到，波及，蔓延到	5A
square grid	方格网，方栅	4A

词 组	意 义	课次
Stacked Bar Code	层排式二维条码	5B
stand for	代表，表示；代替	2A
standardization body	标准化组织	8A
star network	星型网络	8A
stereo system	立体音响系统	7B
storage space	存储空间	10A
stream data	流数据	11B
streaming ads	流广告	5B
study group	学习研讨会，研究小组	9A
subway pass	地铁通行卡	6B
supply chain	供应链	1B
switch over	转变	10B
symmetric cryptography	对称加密	9A
system loss	系统损耗	
take advantage of	利用	4B
take care of	承担	10A
take into account	重视，考虑	1A
take off	腾飞，突然成功	5B
take up	占用	10A
Tamed Frequency Modulation (TFM)	平滑调频	
tamper resistance	抗干扰	3B
tandem switching	汇接交换	
tap into	挖掘	10A
tape recorder	磁带录音机	4A
target server	目标服务器	2A
telecommunications company	电信公司，通信公司	4B
theft protection	防盗	1B
thin film	薄膜	
Thin Film Transistors (TFTs)	薄膜晶体管	3A
think outside of the box	跳出时间框框，创意思维，打破常规	4A
Time Division Multiple Access (TDMA)	时分多址	
time frame	时帧	
time slot	时间空档，时隙	9A
timeout-based retransmission	基于超时重传，基于超时重发	9A
tip of the iceberg	露出水面的冰山顶，事物的表面部分	11B

词 组	意 义	课次
to some extent	在某种程度上，(多少)有一点	6B
token counter	收银台，柜台	6B
Token Ring	令牌网	2B
top out	达到极限	5B
transfer rate	传输率	9A
transmission channel	传输信道	
transmission line	传输线	2A
transmission loss	传输损耗	
transmission path	传输通道	
transport protocol	传输协议	4B
trash can	垃圾桶，垃圾箱	6B
trigger action	触发作用	1A
trunked base station	集群基站	
tunnel interface	隧道接口	4B
ubiquitous networking system	泛在网系统	4A
ultrasonic location system	超声波定位系统	4A
ultrasonic signal detector	超声检测信号探测器	4A
ultrasonic transmitter	超声发射机，超声波发射器	4A
under certain circumstance	在某种情况下	9A
under lock and key	妥善保管，妥善存放	10A
under the guise of	假借，以……为幌子	1A
underlying structure	底层结构	9B
unfamiliar with	不熟悉	5B
unidirectional control	单向控制	
United States Federal Communications Commission	美国联邦通信委员会	10B
urban agglomeration	都市聚集	11A
user interface	用户界面，用户接口	11A
value chain	价值链	1A
video game	视频游戏	10A
virtual memory	虚拟内存	8A
virtual network	虚拟网络	
volatile memory	非永久性存储器，易失存储器	3A
wake up	醒来，唤醒	6A
walkie talkie	携带式无线电话机，步话机	7A
waste management	废物管理	11A

词 组	意 义	课次
waste product	废品，次品	1A
water supply	给水，自来水，供水系统	11A
water tank	水箱，水槽	8B
web page	网页	5A
Wide Area Network (WAN)	广域网	2B
wind up with …	以……结束	6A
Wireless Embedded Internet	无线嵌入式因特网	9A
wireless light switch	无线光交换机	9B
wireless network	无线网络	7A
Wireless Sensor Network (WSN)	无线传感器网络	8A
with age	随着年龄的增长，因年久	6A
word processing	字处理	10A

缩 写 表

缩 写	意 义	课次
4G (Fourth Generation)	第四代移动通信及其技术	10B
8DPSK (Differential encoded 8-ary Phase Shift Keying)	差分编码八元相移键控	7B
AIDC (Automatic Identification and Data Capture)	自动识别与数据捕捉	5A
ARM (Advanced RISC Machines)	公司名	8A
ASIC (Application Specific Integrated Circuit)	特定用途集成电路	8A
ATM (Asynchronous Transfer Mode)	异步传输模式	2B
ATM (Automatic Teller Machine)	自动取款(出纳)机	6A
BPSK (Binary Phase Shift Keying)	二进制移相键控	9A
BSE (Bovino Spongiform Encephalopathy)	牛绵状脑病	1B
CASAGRAS (Coordination and Support Action for Global RFID-related Activities and Standardization)	全球RFID运作及标准化协调支持行动	1A
CCD (Charge Coupled Device)	电荷耦合器件	5A
CCK (Complementary Code Keying)	补码键控	7A
CERP (Cluster of European Research Projects)	欧盟物联网研究项目组	1A
CMOS (Complementary Metal Oxide Semiconductor)	互补金属氧化物半导体	3A

缩 写	意 义	课次
CSMA (Carrier Sense Multiple Access)	载波侦听多路访问	9B
CSMA/CA (Carrier Sense Multiple Access with Collision Avoidance)	载波侦听多点访问/避免冲突	9A
dbm	毫瓦分贝	7B
DC (Direct Current)	直流电	3A
DC (Distribution Center)	配送中心	5A
DECT (Digital Enhanced Cordless Telecommunications)	数字增强无绳电话	7B
DNS (Domain Name Server)	域名服务器	2A
DQPSK (Differential Quadrature Reference Phase Shift Keying)	差分四相相移键控	7B
DSL (Digital Subscriber Line)	数字用户线路	7A
DUN (Dial-Up Networking)	拨号网络应用	7B
EDI (Electronic Data Interchange)	电子数据交换	5A
EDR (Enhanced Data Rate)	增强数据率	7B
EEPROM (Electrically Erasable Programmable Read-Only Memory)	电可擦除只读存储器	3A
EHF (Extremely High Frequency)	极高频	3A
EPOSS (the European Technology Platform on Smart Systems Integration)	欧洲智能系统集成技术平台	1A
ESP (Enterprise Service Provider)	企业服务提供商	4B
EU (European Union)	欧盟	11A
FRAM(Ferroelectric RAM)	铁电存储器	3A
FTP (File Transfer Protocol)	文件传输协议	2A
GFSK (Gaussian Frequency Shift Keying)	高斯移频键控	9A
GTS (Guaranteed Time Slot)	保障时隙	9B
ART (Highway Addressable Remote Transducer Protocol)	可寻址远程传感器高速通道的开放通信协议	8A
HD (High Definition)	高清晰度	10B
HDTV (High Definition TV)	高清晰度电视	10B
HF (High Frequency)	高频	3A
HTML (Hyper Text Markup Language)	超文本标识语言	6B
IC (Integrate Circuit)	集成电路	3A
ICANN (Internet Corporation for Assigned Names and Numbers)	互联网名称与数字地址分配机构	3B
ICT (Information and Communication Technology)	信息和通信技术	11A

缩 写	意 义	课次
ID (IDentification，IDentity)	身份	3B
IEEE (Institute of Electrical and Electronics Engineers)	电气和电子工程师协会	8A
IETF (Internet Engineering Task Force)	因特网工程工作小组	2A
IMT (International Mobile Telecommunications)	国际移动通信	10B
IP (Internet Protocol)	网际协议	11A
IP (Internet Protocol)	因特网协议	2A
ISA (Industry Standard Architecture)	工业标准结构	8A
ISDN (Integrated Services Digital Network)	综合业务服务网，综合服务数字网	4B
ISM (Industrial，Scientific and Medical)	工业，科学和医学	7B
ISP (Internet Service Provider)	因特网服务提供商	2A
IT (Information Technology)	信息技术	4B
ITU (International Telecommunications Union)	国际电信联盟	10B
ITU-R (Radio communication Sector of ITU)	国际电信联盟无线电通信部门	10B
kbps (kilobits per second)	千位/秒	9B
LED (Light-Emitting Diode)	发光二极管	4A
MAC (Media Access Control)	介质访问控制，媒体存取控制	8A
MEMS (Micro-ElectroMechanical Systems)	微型机电系统(微机电)	3A
MiWi	Microchip 无线协议	9A
Mph (Megabits per second)	每秒兆字节	10B
MPSK (Mary Phase Shift Keying)	多进制相移键控	9A
MQTT (Message Queue Telemetry Transport)	消息队列遥测传输	11B
msec (millisecond)	毫秒	9B
MSPs (Managed Service Providers)	托管服务提供者	11B
NAS (Network Access Server)	网络访问服务器	4B
NFC (Near Field Communication)	近场通信，近距离无线通信技术	1B
OBEX (OBject Exchange)	对象交换	7B
OFDM (Orthogonal Frequency-Division Multiplexing)	正交频分复用	7A
O-QPSK (offset Quadrature Phase Shift Keying)	偏移四相相移键控，偏移正交相移键控	9A
OSI (Open Systems Interconnection)	开放式系统互联参考模型	2B
P2P (Peer-to-Peer)	点对点技术	8A
PaaS (Platform as a Service)	平台即服务	11B
PCI (Peripheral Component Interconnect)	外设部件互连标准	7A
PDA (Personal Digital Assistant)	个人数字助理	4A

缩 写	意 义	课次
PIN (Personal Identification Number)	个人身份号码	6A
PKI (Public Key Infrastructure)	公钥基础设施	11B
PML (Product Markup Language)	产品标识语言	6B
RAS (Remote-Access Server)	远程访问服务器	4B
RF (Radio Frequency)	射频，无线电频率	3A
RFID (Radio-Frequency IDentification)	射频识别	1A
RFID (Radio-Frequency IDentification)	射频识别	6A
SDN (Software-Defined Networking)	软件定义网络	11B
SHF (Super High Frequency)	特高频	3A
SOA (Service-Oriented Architecture)	面向服务的体系结构	1A
TCP (Transmission Control Protocol)	传输控制协议	2A
TDMA (Time Division Multiple Access)	时分多址访问	6B
TKIP (Temporal Key Integrity Protocol)	暂时密钥集成协议，临时密钥完整协议	7A
UHF (Ultra High Frequency)	超高频	3A
UID (Unique Identifying Number)	唯一识别码	5A
UID (User Identifier)	用户名	3B
UPC (Universal Product Code)	通用产品码	5A
UPS (United Parcel Service)	美国联合包裹公司	5B
URI (Uniform Resource Identifier)	统一资源标识符	1A,3B
USB (Universal Serial Bus)	通用串行总线	7A,5A
VNC (Virtual Network Computing)	虚拟网络计算	4A
VPDN (Virtual Private Dial-up Network)	虚拟专用拨号网	4B
VPN (Virtual Private Network)	虚拟专用网络，虚拟个人网络	4B
WAN (Wide Area Network)	广域网	7A
WEP (Wired Equivalency Privacy)	有线对等隐私，有线对等加密	7A
WiFi (Wireless Fidelity)	无线局域网	7A
WPA (WiFi Protected Access)	无线安全访问	7A
WORM (Write Once, Read Many)	单次写多次读	6B
XML (eXtensible Markup Language)	可扩展标识语言	3A